© Mehran Mohajer

About the Author

CHRISTOPHER DE BELLAIGUE was born in London in 1971 and has spent the past decade in the Middle East and South Asia. He writes for the *Economist*, the *New York Review of Books*, *Granta*, the *New Yorker*, and the *New York Times Book Review*. He lives in Tehran with his wife and son.

IN THE ROSE GARDEN
OF THE MARTYRS

IN THE ROSE GARDEN
OF THE MARTYRS

A Memoir of Iran

❖ ❖ ❖

CHRISTOPHER DE BELLAIGUE

HARPER ● PERENNIAL

NEW YORK • LONDON • TORONTO • SYDNEY

HARPER ● PERENNIAL

First published in Great Britain in 2004 by HarperCollins Publishers.

A hardcover edition of this book was published in the U.S. in 2005 by HarperCollins Publishers.

FIRST HARPER PERENNIAL EDITION PUBLISHED 2006.

The Library of Congress has catalogued the hardcover edition as follows:
    De Bellaigue, Christopher.
    In the rose garden of the martyrs : a memoir of Iran / Christopher de Bellaigue.
        p. cm.
    Includes bibliographical references.
    ISBN 0-06-620980-3
    1. Iran—Description and travel. 2. Iran—Social life and customs—20th century. 3. De Bellaigue, Christopher.—Travel—Iran. I. Title.
    DS259.2.D43   2005
    955.05'42—dc22   2004059655

ISBN-10: 0-06-093536-7 (pbk.)
ISBN-13: 978-0-06-093536-8 (pbk.)

06 07 08 09 10 ❖/RRD 10 9 8 7 6 5 4 3 2 1

Each day more than yesterday,
and less than tomorrow

AUTHOR'S NOTE

I have used the Persian transliteration for Arab names when they are found in an Iranian context, such as the mourning ceremonies in Tehran for the Imam Hossein.

CONTENTS

ACKNOWLEDGEMENTS

I am enormously indebted to the men and women who are the main characters of this book. They trustingly described to me the ecstatic and excruciating moments of their lives. Some carry their real names, but the identities of most have been veiled by pseudonyms. A few characters, like the Iraqi soldier, have been created from more than one person.

Many helped on the way. In the early 1990s Charles Melville and the late John Cooper taught me Persian Studies at Cambridge. In Tehran I benefited from the insight and industry of my assistant, Mohsen Asgari. In Isfahan, I learned a great deal from Ahmad Akbar al-Sadat and Moussa Darakshandeh; in Qom, from Muhammad-Hossein Zeynali and Ghollam-Hossein Nadi. My appreciation of Iranian history and politics was greatly enhanced by conversations that I had with Reza Alavi, Saeed Leylaz, Muhammad Paki, Hossein Rassam and Hassan Tehrani. When researching the serial murders, I was greatly helped by Ali-Reza Alavi Tabar, Shadi Sadr, Nasser Zarafshan and Mehran Mohajer.

In Tehran, Jim Muir was a patient landlord. He, Guy Dimore, Serge Michel and Thomas Loudon were stimulating office colleagues. The late Kaveh Golestan was a dynamic if sporadic mentor. Roxanna Shapour was a demanding early reader. Reza Alavi and Hossein Rassam were kind enough to offer their suggestions on the finished manuscript. At the Ministry of Culture and Islamic Guidance, Muhammad-Hossein Khoshvaght and his team, notably Ali-Reza Shiravi and Efat al-Sadat Eqbali-Namin, were unfailingly courteous and helpful. To Camilla Campbell, Jo Weinberg and Leo Amery, I owe thanks for advice and support. Throughout, I benefited from the wisdom and friendship of Barbara Smith, my editor at *The Economist*. My agent, David Godwin, and my editor at HarperCollins, Michael Fishwick, deserve many thanks for putting their faith in me and in this book.

In London, my father, Eric de Bellaigue, provided unstinting support. In Tehran, I owe boundless thanks to my wife, Bita Ghezelayagh, and to her superb family: Abtin, Manijeh, Babak, Amir Ali and Razavieh Khanom.

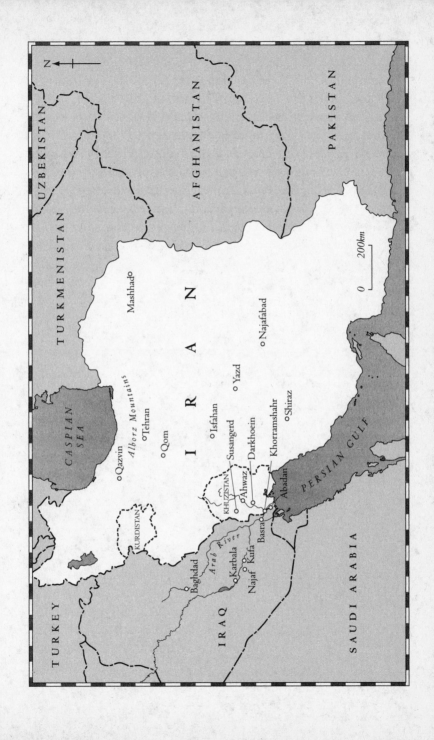

IN THE ROSE GARDEN
OF THE MARTYRS

CHAPTER ONE
Karbala

Why, I wondered long ago, don't the Iranians smile? Even before I first thought of visiting Iran, I remember seeing photographs of thousands of crying Iranians, men and women wearing black. In Iran, I read, laughing in a public place is considered coarse and improper. Later, when I took an oriental studies course at university, I learned that the Islamic Republic of Iran built much of its ideology on the public's longing for a man who died more than thirteen hundred years ago. This is the Imam Hossein, the supreme martyr of Shi'a Islam and a man whose virtue and bravery provide a moral shelter for all. Now that I'm living in Tehran, witness to the interminable sorrow of Iranians for their Imam, I sense that I'm among a people that enjoys grief, relishes it. Iran mourns on a fragrant spring day, while watching a ladybird scale a blade of grass, while making love. This was the case fifty years ago, long before the setting up of the Islamic Republic, and will be the case fifty years hence, after it has gone.

The first time I observed the mourning ceremonies for the Imam Hossein, I was reminded of the Christian penitents of the Middle Ages, dragging crosses through the dust and bringing down whips across their backs. In modern Iran, too, there is self-flagellation and the lifting of heavy things – sometimes a massive timber tabernacle to represent Hossein's bier – as an expression of religious fervour. The Christian penitents were self-serving; calamities such as the Black Death provoked a desire to atone, to save oneself and one's loved ones from divine retribution. Iran's grieving does not have this logic. This is no act of atonement, but a sentimental memorial. Iranians weep for Hossein with gratuitous intimacy. They luxuriate in regret – as if, by living a few extra years, the Imam might have enabled them to negotiate the morass of their own lives. They lick their lips, savour their misfortune.

I see Hossein alongside Tehran's freeways, his name picked out in flowers that have been planted on sheer green verges. I see his picture on the walls of shops and petrol stations, printed on the black cloths that are pinned to the walls of streets. The conventional renderings show a superman with a broad, honest forehead and eyes that are springs of fortitude and compassion. A luxuriant beard attests to Hossein's virility, but his skin is radiant like that of a Hindu goddess. He wears a fine helmet, with a green plume for Islam, and holds a lance. I once asked an elderly Iranian woman to describe Hossein's calamitous death. She spoke as if she had been an eyewitness to it, effortlessly recalling every expression, every word, every doom-laden action. She listed the women and children in Hossein's entourage as if they were members of her own family. She wept her way through half a dozen Kleenexes.

Every Iranian dreams of going to the town of Karbala, the arid shrine in central Iraq that was built at the place where Hossein was martyred. I went there myself, the camp follower of American invaders, and visited the Imam's tomb. Inside a gold plated dome, Iraqis calmly circumambulated a sarcophagus whose silver panels had been worn down from the caress of lips and fingers. They muttered prayers, supplications, remonstrations. Suddenly, the peace was shattered by moans and the pounding of chests, splintered sounds of distress and emotion. Five or six distraught men had approached the sarcophagus. One of them was half collapsed, his hand stretched towards the Imam; the others shoved and slipped like landlubbers on a pitching deck. My Iraqi companion curled his lip in distaste at the melodrama. 'Iranian pilgrims,' he said.

It all goes back to AD 632, when the Prophet Muhammad died and Ali, his cousin and son-in-law, was beaten to the caliphate, first by Abu Bakr, the Prophet's father-in-law, and then by Abu Bakr's successors, Omar and Osman. Ali gave up political and military office, and waited his turn, and the modesty and piety of the Prophet's time was supplanted, according to some historians, by venality and hedonism. After twenty-five years, following Osman's brutal murder, Ali was finally elected to the caliphate. But his rule, although virtuous, lasted only until his murder five years later and gave rise to a rift between his followers and Osman's clan, the Omayyids. The origin of the rift was a dynastic dispute, between supporters of the Prophet's family, represented by Ali, and the Prophet's companions, represented by the first three caliphs. It prefigured a rift that continues, between the Shi'as – literally, the 'partisans of Ali' – and the Sunnis, the followers of the *Sunnah*, the tradition of Muhammad.

After Ali's murder, Hassan, his indolent elder son, struck a deal with the Omayyids. In AD 680 Hassan died and Ali's younger son, Hossein, took over as head of the Prophet's descendants. Hossein was pious and brave and he revived his family's hereditary claim to leadership over Muslims. This brought him into conflict with Yazid, the Omayyid caliph in Damascus. When the residents of Kufa, near Karbala, asked Hossein to liberate them from Yazid, the Imam went out to claim his birthright, setting in train events that led to his martyrdom.

One night, on the eve of the anniversary of Hossein's death, I put on a borrowed black shirt and took a taxi to a working-class area of south Tehran. The main road where the taxi dropped me was already filling with families and men leading sheep by their forelegs. Cauldrons lay by the side of the road. Everyone wore black; even the little girls wore chadors, an unbuttoned length of black cloth that unflatteringly shrouds the female body. I entered a lane with two-storey brick houses along both sides. There was a crowd at the far end of the street, their backs to us, and their silhouettes were flung across the asphalt. Black bunting had been strung between lampposts. Walking towards the crowd, I fell in step with a middle-aged man who was being followed by his family. I heard him mutter, 'Hossein . . .' He looked shocked and puzzled, as if he'd just received news of the Imam's martyrdom.

At the far end of the street there was a stage marked out by pot plants. In the middle of the stage was a bowl of water, resting on a green cloth. The middle-aged man's wife and daughters went to the opposite side of the stage, where the other women and children were gathered under an awning. His teenage son joined a group of young men with gelled hair on the right. To the left was backstage, and an orchestra that consisted of two *tombak* drums and a trumpet. I stayed on the near side. Suddenly, the men in front of us parted to allow a stream of piss, from a camel trembling bow-legged in the arc lights, to run down the street.

A young trumpeter played a riff and the obscene Damascene appeared stage left. (Everyone recognized Yazid: he wore a cape of red and yellow to accentuate his licentiousness, and he wasn't wearing so much as a scrap of green, the colour of Islam.) His helmet was surmounted by yellow plumes. His fat face was expressionless. After prowling around, he started to shout evil words into the microphone he was holding, which was connected to a loudspeaker that in turn felt as though it was connected directly to my ear.

Although he ruled the lands of Islam in the name of Islam, Yazid was notorious for his depravity. Today, Iranians loathe him as if he were still malignantly alive. They recall the menagerie of unclean animals such as dogs and monkeys that he is believed to have kept at court. They talk disapprovingly of the 'coming and going' – a common euphemism for frenetic sexual activity – for which Damascus was known. It is said that he was as devious as he was deviant.

Perhaps Hossein had reckoned without the deviousness. By the time he and his companions bivouacked at Karbala, near the banks of the Euphrates, the caliph had bribed the inhabitants of Kufa to revoke their support for him. His small force was greatly outnumbered by the army that Shemr, Yazid's commander, had raised. Shemr had cut off Hossein's access to the Euphrates, and Mesopotamia in summer is as hot as hell.

Onstage, the players were relating the entreaties, negotiations and moral dilemmas that preceded Hossein's martyrdom. The women and children in Hossein's entourage were suffering from the heat. Since there were no women onstage, we learned this from a narrator, a slim, alert version of the man playing Yazid – his brother, perhaps. Suddenly, there was activity stage left and Yazid returned. The actor's movements and expression were the same, but now he wore green from head to toe. He had changed character and had become Hossein.

As far as I could make out through the echo and distortion, Hossein was relating the anguish that he felt at his decision to fight to the death.

In return for fealty to Yazid, he and his companions would be spared, but that would mean living in dishonour, indifferent to God's will. Then Hossein's half-brother, Abol Fazl, entered.

The portraits show Abol Fazl to be as god-like as his brother, albeit more windswept. The Abol Fazl before us was shifty and greasy; he would have been convincingly cast as a sheep rustler. He was much shorter than Hossein, whom he clasped repeatedly to his breast as they both wept. Hossein was asking Abol Fazl to fetch water from the river. Both knew that the younger brother stood little chance of surviving his mission.

Abol Fazl leaped onto a mangy grey standing at the side of the street, where the camel had been. (The camel was peripatetic and for hire; it was now appearing on other stages in the neighbourhood.) He steered the horse dexterously around the stage, calming it when its hind legs buckled as it turned on the greasy asphalt. Whenever Abol Fazl approached the awning, the women shrank, while he (holding the microphone in one hand and the reins in the other) declared his love for Hossein and for God. The young men in the audience grinned when the horse broke wind during a break in the music. Their fathers frowned.

The next bit of the story happened offstage. Fighting savagely – I had read this in the books – Abol Fazl reached the riverside. He bent down, cupped his hand and brought some water to his mouth. Then he stopped himself and the water flowed back through his fingers. His sense of chivalry wouldn't allow him to slake his thirst before the women and children had slaked theirs. Having filled his leather water container, he remounted, but was cut down in the subsequent struggle, losing his hands and eyes. He cried out, 'Oh brother, hear my call and come to my aid!' Two arrows were dispatched. One pierced Abol Fazl's water container. The other entered his chest.

Abol Fazl staggered onstage. The pierced flagon was between his teeth. An arrow protruded from his chest. His arms were two very long stumps. The stumps supported two bloody objects, which he dropped for us to see: his hands, sliced off in the fray. The Imam cradled the dying Abol Fazl. The men near me in the audience were beating their chests in time with the *tombak*. The women under the awning rocked inconsolably.

And that was the end of the play. It wasn't time for Hossein to die; that would come tomorrow, the day that is called Ashura. The actors picked themselves up and left the stage. Among the audience, there was a rustling, a rearranging of positions and a collective, audible exhalation. And then, to my surprise, the inconsolable found consolation. Facial expressions brightened. The audience's agony changed to equanimity, even satisfaction. The man in front of me greeted the person standing next to him agreeably; a few seconds before, both had been blubbing like children. In the women's section, conversations began. Abol Fazl seemed to have been forgotten.

Had he been forgotten? Was this grief deceitful? Not deceitful, I think: simply not exclusive. The emotions in Iran haven't been compart-mentalised. They coexist; they thrive in public. The borders between grief, entertainment and companionship are porous. You can weep buckets, natter with a neighbour and take away memories of a farting nag. Stifled sobs, trembling upper lips – they don't exist here. Emotion may be cheaply expressed, but that doesn't mean the emotions are cheap.

Some members of the audience were starting to leave their places. The narrator strode into the middle of the stage. He addressed us fluently, softly. He craved our indulgence – he wanted to tell a story that would live in our memories. The people moved back to their places and he began.

A few years back, he started, after the troupe had performed the play we'd just seen, he'd been delighted when a man dropped a large sum of money onto the green cloth in the middle of the stage. As he was counting it after the performance, another man had approached and said, 'Excuse me for interfering, but you can't accept that money.'

The narrator had replied: 'Why not? It's a lot of money, and I've got a wife and kids to feed. It pleases God when money is accepted for good work.' The man replied, 'Believe me, sir, you can't accept this money. Yours is Muslim work, and the man who gave you the money is a Christian. He's Armenian.'

The audience was gripped. What a dilemma! What would you do in such a situation? The narrator went on: 'The Armenian chap was driving off when I ran up to him and thrust the money through the open window of his car. I said, "I'm sorry; I can't accept this money. Forgive me, by the soul of the Imam Hossein, I can't accept."'

When he learned why his money had been rejected, the Armenian had switched off the car ignition and said, 'I have something to tell you.

'Recently, I was driving with one of my employees, a Muslim, and the brakes failed as we were coming down from the mountains. There were valleys on both sides, and we were going faster and faster. I called out, "Oh Jesus! Save us!" and tried the brakes again, but they didn't work. I called out a second time, louder, and rammed my foot down on the brakes. Nothing. A third time, I beseeched Jesus to save us. Again, no result.

'Panic-struck, I looked across at my employee. He said quietly, "Call for Abol Fazl." I was having trouble keeping the car on the road. I shouted, "Who's Abol Fazl?" He said, "Sir, time is running out. Call him!" I had nothing to lose, so I shouted, "Save us, Abol Fazl!" and the brakes suddenly worked. We came to a halt just short of a cliff.

'When we got out of the car, I asked my employee if he'd seen a man on the road, as we were braking. He shook his head. I told him that there had been a man wearing green, and that he had no hands.'

The narrator paused. He bowed his head and emitted three sobs. Then he wiped his eyes and his tone became diffident. 'Estimable brothers and sisters, you may wish to express your appreciation, and it doesn't matter how much you put on the green cloth . . .' – he went on to list sundry denominations, all of which were beyond the means of those present. 'No, the amount doesn't matter. But if, during the course of the coming year, you request Abol Fazl's intercession and he doesn't answer, take the matter up with me . . .'

Nudged by their mothers, the little boys and girls came across to our side of the stage, to get money from their fathers. Then they went over to the cloth and knelt down to kiss and touch it – it had an association, however tenuous, with the Imam Hossein, and few in the audience had the means to go to Karbala. They dropped their money. Once the cloth was covered with notes, men appeared holding trays laden with refreshments. They'd been provided, we learned, by a local trader called Mr Naji. His philanthropy would earn him friends in this life and divine favour in the next.

There were cakes and cucumbers laden high on a copper plate, cinnamon-flavoured rice puddings and little stork's bundles containing deep-fried white candies seasoned with rose water. There was a ewer pouring water into plastic cups, a loop of tea from the spout of a kettle. In her determination to get a rice pudding, a woman elbowed me in the face. I escaped from the crowd.

Rubbing my jaw, I walked away into a nearby side street. The piercing notes of the orchestra had been succeeded by a mellow, distant sound. Gradually, it grew closer and I was able to distinguish individual sounds within it: hands striking chests, a tremor of lamentation and the diesel motors on generators that were amplifying the lamentation. The processions had started.

Suddenly, I heard a scramble of words through a loudspeaker, and the boom of a bass drum. I looked back up the humdrum street, with its box-like parked cars and unsanitary smell coming from the drainage channels, and saw an army of mounted men on the brow of a hill. Their lances scintillated in the lamplight as they prepared to charge and meet their doom.

The army turned out to consist of a man carrying an iron standard, along whose considerable length oscillated swords and gargoyles and plumes of different colours. He was followed by two columns of men, marching in time with the base drum, flagellating their backs with chains on short handles – a strike for every ponderous beat. A man held up an unintended cross that was composed of two loudspeakers tied to a pole; they were wired to a microphone held by a wailing man a few paces behind.

I had to squeeze up against the wall to make room for the standard to pass. The bearer was thickset, bulging and tight-lipped in his task. He was bound to his panoply from a buckle on a thick belt around his waist. As he passed, he half-slipped, and the weight of the standard pulled him towards me. Thinking I might get hit, I ducked into a side alley.

Once the man had passed, I followed the procession to the main road, where it entered a string of processions, a dozen or more from different neighbourhoods. They were united, and also in competition with each other. The people along the pavements would decide which procession was biggest, and which had the most impressive standard. Had the flagellants been equipped with one chain or two? (From that, you could gauge the benefactor's generosity.)

There was a mood of sombre recreation. More young men with heavily gelled hair held suggestive conversations with groups of unescorted girls – this being the only night of the year when young women, under the cover of piety, were allowed to roam without a chaperone. Families strolled. Young boys had been dressed in the white Arab robes of little Ali Akbar, Hossein's nephew; Ali Akbar had fought bravely against Shemr's men, before being cut down.

The more I walked, the better I understood the enormous size of this crowd; it extended as far as the eye could see. This main road was fuller, perhaps, than it would be at any other time in the year. The same was true of main roads across Iran; at that moment, tens of millions of people were in the streets. I reflected on the grief, and the entertainment that people made of that grief. Then I remembered another reason for the show: defiance.

The people on the streets were united by their love for the Imam Hossein, his father the Imam Ali and (to a lesser degree) the other ten Imams that are regarded by most Shi'a Muslims as the rightful inheritors of the Prophet's mantle. Shi'as are an overwhelming majority in Iran, but only a small minority across most of the Muslim world. It was the forebears of today's Sunni majority, Yazid and his followers, who rejected the hereditary principle and murdered its exponent, the Imam Hossein. (They also, Shi'as believe, murdered every other Shi'a Imam, apart from the twelfth.) Even now, in the twenty-first century, the Sunnis of neighbouring Pakistan are capable of launching murderous attacks on the Shi'as of that country, shooting up mosques and assassinating prayer leaders. In Saudi Arabia, a Sunni monarchy controls the holy places. Many Sunnis regard the Shi'as as heretics.

And so here, and in other streets across Iran, the people were showing that they would neither be extinguished nor ignored. They were showing, too, that they would not forget that dreadful sin, the murder of the Imam Hossein.

CHAPTER TWO
Isfahan

One afternoon in the spring I set out from the Armenian quarter in the lovely city of Isfahan, towards the Seminary of the Four Gardens. The following day was the anniversary of the investiture of the Imam Ali as the Prophet's successor. The people were in a good mood. They revered Ali for being modest and just, and looked forward to celebrating these qualities by visiting family members, stuffing themselves with *beryan* – a dish that features minced sheep's lungs – and passing judgement on their hosts' new daughter-in-law. They strolled in the mild afternoon sun, mothers and daughters arm in arm (and fathers in their wake), buying tulips to put in iced water to keep overnight, and sweetmeats to take as gifts.

I reached one of the main roads that head north towards the river, and hailed an old shared taxi. The back seat had its complement of three. The occupant of the front passenger seat stepped out so that I could sit between him and the driver; I was suspended over the gap between their seats. The driver sat hunched over the steering wheel, leaning slightly against the door. We moved off. The driver changed gears like a surgeon replacing dislocated bones.

We were soon stuck in traffic outside one of the big banks, in front of which was a shiny blue car mounted on a gantry. The car – new, French-made – was an incentive: every account holder stood a chance of winning it in a prize draw. It was caparisoned with bunting and flashing light bulbs.

It had metallic paint that had been devised by a computer. The bank had put it on the gantry to publicize it – and to make it hard to steal.

I looked in the rear-view mirror and my eye was taken by a fat woman sitting in the middle of the back seat. She was staring longingly out of the window at the zippy French car. She caught me looking at her and pretended to be scandalized, tucking her fringe under her headscarf. 'What's happening up there, Mr Driver?' she demanded. 'Why aren't we moving?'

A car, a Buick from the 1970s, was stuck at the intersection, having carried out half a U-turn. Another car, an Iranian-made Paykan, had grazed one of the Buick's tailfins. The drivers had got out of their cars. The wife of the Buick driver was leaning out of the window, yelling.

'Look at the wife, egging him on!' said our driver. 'What difference does it make? That poor Buick's been wounded more times than I have.' The side of the Buick was discoloured from dents that had been amateurishly smoothed out. The engine was still running. It emitted black smoke.

The taxi driver reached under his seat, pulled out a thermos and unscrewed the cap. He poured a little tea into a dirty glass that rested on the dashboard, swilled it around and poured it out of the window. He filled the glass with tea and, putting it back on the dashboard, closed the thermos and put it back under his seat. Then he held up the glass and said, 'Please go ahead . . .'

He was offering us tea. In such instances, you don't accept. It would be bad form. It's his tea, but he has to offer it. It would be bad form not to. But he'd be put out if someone said, 'Yes, I'd like some of your tea.' No one does. The driver gets to drink his tea and appear courteous at the same time. Both ways he wins.

There was polite murmuring around the taxi: 'Thanks, but no' . . . 'You go ahead and have some' . . . 'I don't feel like tea' . . . 'I've just had some tea.'

Lies. We'd all enjoy a glass of tea.

The driver took out a packet of cigarettes and we went through the same rigmarole. We felt our breast pockets for imaginary packets of cigarettes. Eventually, the driver withdrew a cigarette from his packet, lit it and settled down to watch. A policeman had arrived at the intersection. He was trying to broker a reconciliation. The driver of the Paykan was a cocky brute, well-built, young enough to be the Buick driver's son. He danced from one foot to another. Soon, the policeman seemed to make a breakthrough. The youth hugged the Buick driver.

During the argument, the traffic lights at the intersection had turned green several times, at which cars had surged forward from all directions. Lots of them wanted to turn, this way or that, but the Buick and the Paykan were blocking their way. The cars were revving, edging forward, kissing bumpers. Someone would have to reverse. Iranian drivers don't like reversing. It's a form of defeat. I felt sorry for the policeman.

He did a good job. He positioned himself in the middle – whistling, gesturing, occasionally giving a winning smile. He was a professional. In a little while, at his prompting, a car edged forward from the middle, and away. Another followed. The knot was untied.

'Well done!' the taxi driver murmured, and we moved forward. The protagonists stayed where they had been. They would wait for more policemen, who would take statements and measure angles to determine who was at fault. As we went past, the Buick driver's wife, a woman in a red scarf, leaned out of the window and shouted at her husband, 'I should have known you wouldn't have the balls to stand up for yourself! You, who took the full brunt of the Iraqi attacks! Why don't you stand firm, instead of letting some beardless chick trample your pride?'

The woman's husband turned around. His face was full of anguish. His wife wasn't much older than the Paykan driver.

The taxi driver sighed as we drove off. 'You've got to show them who's boss from day one. I mean, now it's too late. He's let her get out of control, challenge his authority. Nothing he can do now.'

A little further down the road, a man who was sitting next to the woman in the back seat got out. He was replaced by a thin woman who recognized the succulent woman: they were distant relatives. They didn't seem pleased to see one another. They passed on regards to each other's families, and extended invitations for tea and lunch.

The thin woman said, 'Did you get much rain in Tehran?'

'More than dear Isfahan, I can tell you! You know, what with struggling to combat the illness of my late husband – may God show him mercy – and the demands it's made on my time and health, this is the first time

I've been to Isfahan for five years. Oh! My heart burned when I saw the river – dried up like a burned courgette, with the wretched boatmen standing around in the mud, with nothing else to do but pray for rain. I mean, is it possible for a river to have no water? Our river? In this day and age?'

'They sold our water to Yazd,' the driver said. 'They sent it off in a pipeline. Cost a fortune to build. The fathers of bitches.'

We were in a long queue of cars. The driver leaned out, far enough to see past the cars in front. He swung the wheel and pressed down hard on the accelerator. We emerged from the queue of cars, into the oncoming traffic. There weren't many cars coming; the lights ahead were red. By the time the oncoming traffic started to move, we were elbowing our way into a gap between two cars, now much nearer the traffic lights. One of the other drivers raised his hand, but was too lazy to clench it.

'I don't know why everyone drives so fast,' the fat woman said to her relative. 'All they do when they get to their destination is drink tea.'

The driver grinned. 'God forbid, madam, you were offended by my efforts to expedite you to your destination! Or perhaps it was what I said? Do you have Yazdi blood, by any chance?'

'Lord, no! My parents – may God show them mercy – were from Isfahan, and proud of it. But the president is from Yazd, isn't he?' she said slyly. 'That might explain why they're allowed to drink our water. The Yazdis have always had it in for Isfahan. I should know; my son married a Yazdi. She won't even iron his shirts. She says he gets through too many. He gives them to me, my poor darling. Too proud to iron an Isfahani's white shirt, the Yazdis are!'

'At least they opened the dam again, in time for the holidays,' said the third passenger in the back seat. 'There's water in the river now, thanks be to God.'

'Exactly!' said the fat woman. 'They were scared the Isfahanis would flay them if they didn't open the sluices. But they'll shut the dam again after the holiday, and say there's no more water. They'll send it to Yazd instead.'

'And our poor Isfahani kids will carry on topping themselves,' the man said. 'Everyone knows the suicide rate goes up when the river's dry. It's bad for the soul.'

The man next to me stirred in his seat. 'Pardon me, but you're wrong. The problem is not Yazd, but the farmers in Isfahan province. They're planting rice along the river banks, even though rice needs more water than almost any other crop. Only an idiot would plant rice when there's a drought.'

'And what would you have us eat if there's no rice?' the fat woman demanded. 'You want us to get thin and weak?'

'We should buy our rice from elsewhere.'

'Sir, you'd prefer that we eat Pakistani rice that has no perfume? Or that sticky revolting stuff the Turks call rice? You can't make a respectable *polov* with that.'

The man sitting next to her said, 'She's right; our rice is the best in the world. Everyone says so.'

'And there's another thing,' said the woman, 'our dear motherland has been dependent on foreigners for hundreds of years. Now you want to put our bellies at the mercy of Pakistan! Everyone knows who's behind Pakistan: the English! It wouldn't surprise me if the English had something to do with our water shortage. They always stir up trouble in countries they fear. That's why they're the best politicians, and we've never been any good.'

'The English are indeed very devious,' said the man next to me, 'but I haven't heard of them altering the climate.'

The woman snorted. 'I wouldn't put anything past them.' Then she said, 'With your permission, Mr Driver, I'll get out here.'

The thin woman said, 'I thought your brother lived further on.'

'He does,' the fat woman replied. 'But I like to exercise before a holiday. I'll walk the last half-kilometre.' The taxi stopped. The thin woman got out to allow the fat woman to do so. The fat woman put out both her arms to try and lever herself from the hollow she had created in the back seat. For a moment, one of her hot hands gripped my shoulder. She stood at the window, and looked in.*

---

* You should know about *ta'aruf*. In Arabic *ta'aruf* means behaviour that is appropriate and customary; in Iran, it has been corrupted and denotes ceremonial insincerity. Not in a pejorative sense; Iran is the only country I know where hypocrisy is prized as a social and commercial skill.

Three examples:

When the taxi driver offered us tea and cigarettes, and we refused, this was *ta'aruf*. He

The fat woman said: 'How much, sir?'

'Be my guest,' said the driver.

The fat woman said: 'I beg of you.'

'Whatever you like,' he grinned. 'Really, it's not important.'

'How much? I beg of you.' The woman was getting out her purse.

'I'm serious; be my guest.'

'How much?'

The driver surrendered. 'Seventy-five *tomans*, if you'd be so kind.'

'Seventy-five *tomans*? I only got in at Hakim Street. It's fifty *tomans* from there.'

The driver frowned. 'Seventy-five. It's been seventy-five *tomans* for three weeks now.'

'I gave fifty *tomans* two days ago. I'm not giving more than fifty.' She looked sharply at her relative who was examining her nails.

'It's seventy-five *tomans*,' said the driver. His smile had disappeared.

Suddenly, the woman was angry. 'Is this the correct treatment, the day before we celebrate the investiture of the Imam Ali, *salaam* to him and his family?' She looked accusingly at me. 'Is this the right impression to give foreigners, that Iran's a country of unprincipled hat-lifters? I'm not giving a penny more than fifty.' She threw the note in the window.

The driver picked it off my knee. As he put the car into gear, he said, 'She eats my head with her worthless prattle. She's too stingy to stay in as far as her destination. Then, she rips me off.'

'We're only related by marriage,' said the thin woman.

I said: 'I may as well get out here, Mr Driver. I want to cross the bridge.'

---

had no intention of giving us tea and cigarettes, and we reacted accordingly. A man may propose that his son marry the daughter of his impoverished younger brother without having any intention of permitting the match; the son is already engaged to the daughter of an ayatollah, and the brother's daughter is a repulsive dwarf. But the quintessence of *ta'aruf* can be found in the behaviour of a mullah I once observed entering a Tehran hospital in the company of several other men. As the mullah crossed the threshold, he said to the men waiting behind him, 'After you.'

If, through some mistake or misunderstanding, an offer extended through *ta'aruf* is accepted, it will be retroactively countermanded. I remember reading somewhere of a foreigner who was arrested for theft after being denounced by a shopkeeper who had repeatedly refused to take his money.

'Where are you from?' said the driver, as I gave him the fare.

'France,' I said.

He patted my shoulder. 'Whatever you do, don't marry an Iranian.'

❖ ❖ ❖

I entered the bridge of Allahvardi Khan. Framed in one of the pierced arches was a middle-aged couple, staring at each other. I touched the bricks. They were warm and biscuity. When I reached the other side, I looked back. The Islamic arch had been repeated like the name of God in a prayer.

In the first years of the seventeenth century, these bricks were baking in the name of Shah Abbas I, castles of them hardening over smoking dung. Between 1598 – when Abbas moved his capital to Isfahan from the northern city of Ghazvin – and his death in 1629, they turned a provincial town into one of the world's most opulent capitals.

By moving to Isfahan, Abbas changed the nature of a country whose extremities now roughly corresponded to the borders of modern Iran. (At its peak, his empire encompassed the Iranian plateau, with fingers reaching into Mesopotamia and Anatolia to the west, into the Caucasus to the northwest, and almost to the River Oxus, the northern boundary of modern Afghanistan, in the northeast). Rather than stay near the Caspian Sea, as his Turkmen ancestors had done, Abbas aimed at the centre.

The migration allowed Abbas to give up his former dependence on Turkmen tribesmen, and to set up a new confederation. His government and army contained not only Persians, Turkmens and Arabs, but also Georgian, Caucasian and Circassian converts to Islam. He forcibly imported three thousand Armenian Christian families to Isfahan, and encouraged them to prosper spiritually as well as economically. Foreign visitors found in Isfahan a suitable seat for a cosmopolitan empire – Ghazvin, by comparison, had been a draughty Turkish tent.

Abbas enjoyed the company of foreigners. They, confused by the name of his dynasty, *Safavi*, called him the Sophy. Like his near-contemporary, India's Akbar, Abbas discussed religious questions with the Augustinians and the Carmelites. Like Akbar, he resisted their efforts to convert him.

The Balenciagas, Faberges and Dunhills of the age spoke Persian. During

Abbas's reign, Europe acquired a taste for Persian goods – for silken carpets brocaded with silver and gold, damasks and taffetas, bezoar stones and turquoises. They learned to trip on Persian opium. Abbas's wealth was axiomatic; Fabian wouldn't stop baiting poor Malvolio even 'for a pension of thousands to be paid from the Sophy'.

Abbas was not a successful family man. He murdered his eldest son, Mirza, and blinded the second, Khodabandeh – ruling him out, according to Islamic law, of the succession. Jane Dieulafoy, a formidably disapproving French archaeologist and traveller of the nineteenth century, relates an account she heard of Khoda Bendeh's revenge – apparently exacted on his own small daughter, in order to spite Abbas, who adored his grandchildren:

> One morning, at the very moment when the child came to kiss his unseeing pupils, he seized her and slit her throat, in full view of his panic-struck wife. Then, he threw himself on his son, who had come running at the sound of the struggle, and tried to deal him the same fate. In vain; the child was snatched – still alive – from his father's arms, and Shah Abbas was informed of what had happened. When he was confronted by the corpse of his granddaughter, the old king emitted exclamations of rage and desperation that filled the killer with an exultant and dastardly happiness; for a few moments, he savoured his horrendous revenge, before ending his own life by swallowing poison.

Abbas's fear of his sons perhaps kept him alive; it also prevented promising princes from maturing into worthy rulers. Most of the Safavid Shahs who came after Abbas rivalled themselves only for despotism and sloth. For the remainder of the seventeenth century and the first quarter of the eighteenth, the empire was defended only by one or two competent grand viziers, and the structural excellence of Abbas's state.

Today, Abbas's paranoia has been forgiven. Even in a regime that hates and fears monarchs, people refer to him as Abbas the Great. Hard-line revolutionaries concede his achievements – though they are loath to admit that, were it not for him, their revolution could not have happened. Not only did Abbas help set the boundaries that delineate modern Iran, he also made Iran institutionally, irrevocably, a Shi'a state.

His uncle, the mystic Ismail, had imposed Shi'ism on Iran's mostly Sunni population. But many orthodox Shi'as considered Ismail to be a heretic. His self-depiction as (variously) the harbinger of the twelfth Imam, the twelfth Imam, the Imam Ali, even God, drew to him deluded fanatics who believed he was immortal and impossible to defeat. (Until, that is, his army was smashed by the Turks.) His poetry was denounced as blasphemous. Even by the standards of the time, he drank and sexed immoderately.

Abbas was more conventional – and more inscrutable – than his uncle. He was tempted by flesh and wine, but he dropped Ismail's claims to divinity. His zeal, though sincere, was complemented by his politics; his promotion of Shi'ism as a state religion helped set Iran apart from two predatory Sunni empires in the vicinity: the Ottomans and the Mughals of north India. One of his most important acts was to promote orthodox Shi'a clerics. State-sponsored mullahs were expected to be loyal and to counter the influence of mysticism. (They had a personal interest in doing so. Mysticism's emphasis on the believer's personal relationship with God undermines the mullahs' perception of Islam as primarily a code of laws and behaviour, belittling the transmitters of that code – the mullahs themselves.) Abbas endowed Shi'a seminaries that attracted clerics from other Shi'a centres, like Bahrain and southern Lebanon; he himself married the daughter of one of these foreign clerics.

Scholars in the seminaries learned to understand and interpret Islamic law – through logic, grammar and rhetoric. They learned the relationship between Islamic law and their sources, the sayings and traditions of the Prophet Muhammad and the twelve Imams. They were taught a set of systematic principles for deriving one from the other, called jurisprudence. In time, senior mullahs started issuing new and comprehensive compilations of the sources.

Islam has no sacrament requiring ordained ministers; there is, strictly speaking, no 'clergy' – certainly not in the sense of a homogeneous group of professionals whose job is to mediate between people and God. In the Safavid period, however, Iran gained a clergy in all but name, and it became a social and political institution. Experienced mullahs were sent to the provinces as judges, dispensing Islamic law. They administered wealthy religious foundations. They systematized the collection of religious taxes that entered their own coffers. They became the state's spiritual backer.

The expanding science of jurisprudence legitimized their influence. Jurisprudence allowed senior clerics to interpret religious rulings. The most senior of the jurists – the *mojtahed* – was deemed qualified to divine God's will in areas where he had not expressed himself; this made the *mojtahed* a kind of divine legislator. As the Safavid era wore on, the Shah ceased in religious terms to be more than the titular head of Ismail's old mystic order. He came to rely on the *mojtahed* for religious sanction of his policies and actions. The Safavid-era mullahs did not go as far as to demand political leadership; but that did not stop some of them acquiring a taste for worldly power.

Shah Sultan Hossein, Abbas's great-great-great grandson, came under the influence of mullahs who persuaded him to forbid alcoholic revels and to banish mystics from the capital. He endowed the Seminary of the Four Gardens in Isfahan, to propagate the theology of these mullahs. He authorized the persecution and forcible conversion of Sunnis under his control, as well as minorities such as Christians, Jews and Zoroastrians.

Sultan Hossein's bigotry, combined with indecision and misrule, led to revolt. In 1722, an army of Sunni Afghans captured Isfahan. After keeping Sultan Hossein captive for a few years, they executed him, effectively extinguishing the Safavid dynasty. Iran sank into anarchy and the clergy withdrew from sight.

I walked up the Four Gardens. It had once been four recreational gardens that were laid out by Abbas, with arcades made up of plane trees bowing to one another and a track for horsemen. Now it's a straight, modern road, with travel agents and cake shops. After about half a kilometre I came to a wall of arch shapes illuminated by tiles – the Seminary of the Four Gardens, Sultan Hossein's endowment. I pushed open a door and went in.

After the movement and noise of the street, the seminary gave me an immense sensation of peace. It was laid around a courtyard, bounded by cells set in vaulted niches, with tiled porticos on three sides. A rectangular pool of water and a path divided the grass into four lawns. The cypress trees almost obscured the vivid blue dome over the prayer hall.

A mullah strolled along the pool of water, talking to a seminarian. When they reached the far end, they turned around and retraced their steps. Other seminarians were crossing the courtyard, on their way to class. A few were sunning themselves on the balconies of the first-floor cells. A door slammed, the way they do in institutions.

I walked to one of the corners of the courtyard. Its arch led into a roofless chamber with low stone platforms; in the old days, the mullahs would lecture from these platforms, the seminarians at their feet. I sat down in the shade.

❈　❈　❈

Back in the 1970s, Isfahan was sinking under slime. The King Mother's eastern wall kissed Iran's most opulent hotel, the Shah Abbas. (The Shah Abbas had been a traditional travellers' rest house; now, it had a slab of modern rooms stuck on the front, and a kind of unending feast of Balthazaar going on inside.) Outside the door of the seminary, in the Four Gardens, cars blared Western music. Their young occupants lusted for a US college education. Everywhere, there were signs of progress. Advertisements for washing machines; Old Spice aerosols in pharmacy windows; female arms sprouting downy hairs coming out of halter tops. You could buy foreign booze in the Four Gardens and go whoring round the back of the municipality.

The Shah was Muhammad Reza Pahlavi. He hated mullahs almost as much as he hated Communists; the mullahs were the forces of black reaction, sabotaging his attempt to make Iran modern. The King of Kings had put Isfahan's religious foundations in the hands of a retired general. Perhaps the general had visited Notre Dame or the Duomo; he'd certainly heard how Europe was neutralizing its own black reaction by turning churches into museums. Christianity was changing from a religion into a secular way of appreciating beauty. Could Islam undergo a similar lobotomy?

The general threw open the seminary doors. Some of the mullahs protested. They argued that the seminary was an all-male place of study, whose architectural beauty was designed not to delight strangers but to inspire the seminarian. Why, they asked, had the seminaries been built looking in on themselves? (Answer: to protect the religious scholar from worldly temptation and to reflect his harmonious soul.)

Paying their price of entry, the tourists came into the Seminary of the King Mother, wandering around in shorts and Jesus sandals, peering into cell windows, hoping to catch a seminarian at prayer-whirling, perhaps? On hot days, they dangled their feet in the pool. They asked for postcards, ice cream, toilets.

Gradually, the seminarians were driven out. They found it impossible to concentrate on their studies. Some were lured by moral corruption. Rumours abounded of ghosts, restless mullahs from the days of Sultan Hossein, warning of defilement. Some of them took cells in other seminaries, off the tourist track. Their hatred for the Shah expanded; it became contempt for the Western model that he was trying to impose on them.

The tourists had been attracted by Iran's antiquity and culture, and in some cases by the person of the Shah and his succession of lovely wives. The sportsmen and women among them may have seen the King of Kings from a distance – at St Moritz, perhaps, where he kept a chalet and skied beautifully.

The Shah was America's friend. He was the West's bulwark against Communism. You only had to open *Time* magazine to learn that America wouldn't let him fall. As they toured the city, the tourists occasionally solicited the political opinions of a shopkeeper. There were broad smiles. A signed photograph of the Shah with his third wife, the tirelessly charitable Farah, was produced from a drawer.

The tourists were unaware that they and the shopkeepers were being monitored by Savak, the Shah's US-trained secret police. They didn't realize that everyone they came into contact with had been intimidated or bought. They didn't know – perhaps they didn't care to know – about the bastinadoes, the electrodes and the rectal violations that were the speciality of Savak safe houses.

One evening, the tourists gathered in the courtyard of the Hotel Shah Abbas. They raised their glasses to Isfahan's beauty – to the Safavid architecture, to the Armenian and Jewish quarters.

'And to the Shah!' the smiling *maître d'hôtel* interjected.

The tourists were beside themselves. The Shah's picture was in the lobby, and the restaurant, and at the entrance to the swimming pool. But this was different: a spontaneous show of fealty.

'To the Shah!' they cried.

There was a second set of foreigners, drinking in the hotel courtyard. They were based in the capital, Tehran, but sometimes spent the weekend in Isfahan. They were oilmen and arms dealers, petrochemicals salesmen and dam-builders. They had come to Iran to suggest to the Shah ways of disposing of his massive oil revenues. They spent a lot of time and money bribing ministers and bureaucrats, chasing contracts that would allow them to retire. They enjoyed smearing thick-grained Caspian caviar on crustless toast, posing a shard of lemon peel on top and shoving the whole lot into their mouths.

The third group of foreigners was composed of US Air Force officers. They worked as engineers, instructors, communications officers at Iran's biggest air base, outside Isfahan. Every Isfahani girl had a crush on a US Air Force officer. Their brothers dreamed of piloting a Tomcat. In the bazaar, among the butch porters, blond American boys were all the rage.

The Revolution started sometime in the late twentieth century. Who knows when?

The leftists say it started at the party of 1971, when the world's despots, dynasts and democrats dined with the King of Kings at repugnant expense in the ruins of Persepolis, the magnificent temple complex that was started by the Achaemenian King, Darius, in 520 BC.*

---

* I have a book, *Celebration at Persepolis*, that commemorates this party, which was held in celebration of what the Shah arbitrarily judged to be the two thousand five hundredth anniversary of continuous Iranian monarchy. The book relates that some sixty tents the size of villas, designed by a Parisian firm in beige and royal blue, were erected to house the guests, and that a hatter was on hand should one of the guests squash his topper. Haile Selassie brought with him a Chihuahua wearing a diamond-studded collar. A breakfast of raw camel meat was made available for the Arab emirs. The dinner menu included quail eggs stuffed with Caspian caviar, saddle of lamb with truffles and roast peacock stuffed with *foie gras*. The *vin d'honneur* was Château Lafite Rothschild 1945. Representing the Vatican, I learned, was Cardinal Maximilian de Furstenberg, a relation of my Belgian grandmother's. Although he was only a few years older than her, my grandmother always referred to him as Uncle Max, possibly because he worked for the Pope.

The economists say it started with the oil-price hike two years later, when OPEC quintupled the price of oil. It turned the King of Kings into a superstar, beloved of arms dealers and industrial development gurus, and set inflation on its upward trend.

A taxi driver once told me it started when the people saw the Shah drinking alcohol with his foreign guests, and heard the rumour that certain members of his family liked to swim in milk.

Everyone agrees it had started by the time the Shah made his final trip to Washington, in 1978, when he and Jimmy Carter wept in the White House rose garden – not out of love for each other but because of the tear-gas canisters being fired at anti-Shah demonstrators in Pennsylvania Avenue.

Perhaps it started in Isfahan, the day a boy spat in the face of a German woman who was immodestly dressed.

I'm sitting in a basement in Qom that belongs to Mr Zarif. He's smoking his hookah: short sucks and clouds spreading over his face. He doesn't smoke to relax. The in–out helps him concentrate. He talks faster when he's smoking, and he talks pretty fast anyway. He shouldn't smoke, the doctors have made that clear, but he enjoys doing things he's not supposed to – as long as they don't upset God. Mr Zarif is small and balding. He has a big head and a button nose and ironic eyes. He looks like a djinn, with scented smoke wings.

He folds the snake, as etiquette requires, so that the nozzle faces away from me, before handing it across. His wife will be down in a minute, bringing tea and fruit cut into triangles. She'll tut-tut when she sees the hookah, and she'll smile; the pleasure of watching her husband's pleasure is more powerful than the fear that smoking will kill him. (If God has heavenly plans for you, living well beats living long any day.) Then – for this is an enlightened household, with no fanatical segregation of the sexes – she'll join us, stuffing the end of her chador, which is adorned by a field of peonies, between her teeth as she passes around tea. I've known Mr Zarif for several months, and I think of him as a friend. But it's hard, listening as he explains his past, not to feel as though he's talking about someone else.

Perhaps, I think, he's deliberately trying to give the impression that he bears no relation to the Zarif of two decades ago; the present Zarif can analyse dispassionately the actions of the former Zarif. Perhaps it's a way of shoring up regret or bitterness. Or Mr Zarif is trying to be honest. I've been confronted by two Zarifs, so different as to be enemies, and I want to know what makes them one.

'Have I shown you my *nanchiko*?' Mr Zarif leaps to his feet – I've never known anyone rise from a cross-legged position so compactly and elegantly – and runs out of the room. He comes back holding two bits of wood joined by a chain.

'You know how the Japanese invented this?' I shake my head. It looks good for throttling people. 'There was a time when they had a weak and paranoid Emperor who banned the people from bearing arms. So they went to the obvious place: the kitchen! Someone had the idea of joining two rolling pins with a chain.' He limbers up, rolling his shoulders, crouching slightly. 'Of course, I'm out of practice.'

He starts to whip the *nanchiko* in arcs about his body, threatening adversaries from every angle. The *nanchiko* buckles and snaps. One of Mr Zarif's advantages is his low centre of gravity; knock him down and he'll swoon like a top, bob up again. Wham! The *nanchiko* lashing at you, splitting your forehead, breaking your elbow.

You have to discount Mr Zarif's eyes, which have been dappled by hindsight. Back then, they were . . . what? Angry? Crazy?

This much is certain:

The former Zarif would have had no Englishmen in the basement, smoking the hookah. The former Zarif divided the world into friends and enemies, and the outside world was composed almost exclusively of enemies. (Of course, the British; they occupy a privileged position in Iran's demonology. The former Zarif had things to say about us.)

Mrs Zarif comes in with a tray. She piles my plate high with fruit, and then does the same to Mr Zarif's. She teases me about my appetite, which is known to be insufficient and will be the cause of my enfeeblement. Mr Zarif says I'd better be hungry today, because his wife has made *shirin polov.* It's a feast of barberries, crushed pistachios, walnuts and lamb – on a bed of rice.

The front door slams. It's Ali, the Zarifs' ten-year-old son, back from school. Within a minute or two of being greeted by his parents, he's challenged Mr Zarif to climb through the small hatch between the sitting room and the kitchen, through which Mrs Zarif will pass us lunch.

'Of course I can do it,' says Mr Zarif. He looks at me. 'It wouldn't be right, though, with Mr de Bellaigue here.'

'You can't do it,' Ali smiles. 'You'd get stuck.'

Mr Zarif is smiling, but infuriated. 'Of course I can. Is it that I'm too fat, or too old?'

Ali shrugs viciously, as if to say: 'Try.'

'Well, if Mr de Bellaigue gives permission . . .'

Ali: 'You can't do it.'

Mrs Zarif tells her husband not to be so silly. It's not a very elegant thing for a grown man to do, to climb through the hatch at Ali's urging. I tell him not to hold back on my account.

Mr Zarif climbs onto the little table, puts his hands through the window and levers himself up. For a moment, he's caught on the ledge; he's having trouble manoeuvring his legs around and through the window. But his legs aren't long and he eventually gets through, grunting as he goes. Mr Zarif disappears, and we hear him land on the kitchen floor. When he comes back into the sitting room, his face is red and he's triumphant. Mrs Zarif says, 'I'm sure Mr de Bellaigue is impressed.' Ali is climbing over his dad, ruffling his hair.

In another country, at another time, Mr Zarif would have been called a delinquent, a thug, a menace to society.

❖   ❖   ❖

He was brought up in Isfahan, and he set up his first gang in 1978, when he was twelve. He and his friends copied and distributed illicit pamphlets. They pasted flyers and photographs of dissidents onto walls, at night. (Making sure that no one was around to turn them in to Savak.) The following day, as the people walked to work, they'd see Khomeini looking at them. His eyes would demand: 'What have you done for the morally upright and economically downtrodden?' They would accuse: 'Acquiescence to tyranny makes you an accessory!'

The local officials would be embarrassed; they'd phone the police, who would rush to the scene of the crime and start scraping the papers off the walls. 'Quick, boys! The governor's limousine is cruising up the street!'

The principal at Mr Zarif's school hauled him up for daubing 'Death to the Shah' on a wall. Only the intercession of a friend of his father's, a kind gent from the Education Ministry, saved him from Savak.

I ask: 'Did you understand what you were doing, that you were taking part in a revolution? Or was it just a game?'

Mr Zarif smiles, a you-should-know-better-than-to-ask-that smile. Then he says, 'Khomeini.'

Of course, Khomeini! There was something about him that called out, fathered you. It was impossible not to be scared of Khomeini – imagine him staring at you, like a torch shedding black light! He made you ashamed to breathe the same air as the officials of the King of Kings. Waiting for him to come back, willing his return from exile – first from Iraq, later on from France – people called him Master. The Master. A few months before the Revolution, they started calling him the Imam.

During the months that preceded the Revolution, a rhythm was established. There would be an atrocity – helicopter gunships strafing demonstrators in Tehran, for instance. The atrocity would be followed by an emotional, politicized funeral, which would lead to a second atrocity. More mourning and outrage. A funeral, another atrocity, and so on. There was a second, parallel movement: a roller coaster of panicky sackings and appointments, imperial apologies and admonitions, relaxations and crackdowns.

In Isfahan, rumours spread that the masked soldiers putting down the demonstrations were Americans, helped by Israelis. News spread that someone had shot an American who'd tried to enter a mosque without taking off his shoes. The Americans and their families started going home. The newspapers were full of ads for second-hand washing machines.

On 16 January 1979 the King of Kings flew away, with Farah, a great many jewels and a clod of Iranian earth. Two weeks later, Khomeini returned from exile, dismissed the government that the Shah had left behind and announced a provisional administration.

Mr Zarif saw things clearly. This is what he saw:

History had restarted with the Revolution and Khomeini's return from exile – just as it had restarted with the Prophet's migration from Mecca to Medina in AD 622, and the establishment of the first Islamic administration. The Imam would recreate the pure Islamic rule that Muslims had only known under the Prophet and later on, for five years, under the Imam Ali. There would be social justice, for social justice is inherent in Islam. Society would be cleansed of Western influence. Whatever the Imam decreed, that would happen. There was no question of challenging the Imam's authority, for that would be the equivalent of challenging God.

The Revolution would start in Iran, before moving on to the rest of the world. Muslim countries would be first. Islamic revolutionaries would sweep away the house of Saud and Turkey's despotic secularism. They would liberate Iraq from the pseudo-Socialism of the Baath Party, and restore Iraq's oppressed Shi'a majority to their rightful position of dominance. A column of revolutionaries, led by Iranians, would march into Jerusalem and say their prayers at the al-Aqsa Mosque. Israel would be destroyed, although some Jews would be allowed to stay on. (The Qoran makes provision for the coexistence of Jews, Christians and Muslims, so long as the Jews and Christians accept their inferior status.)

Not everyone saw things as clearly as Mr Zarif. You only had to look at the provisional government to realize that the Imam had been forced to share power with undesirables. Many in the government saw the future through a kaleidoscope that had been manufactured in the West. They defined Islam in Western terms. They shouted the same slogans as the ideologues, but they meant different things.

Take the prime minister, Mehdi Bazargan. Although Bazargan was personally pious, he was a professed 'democrat'. He wore broad ties. Soon after Khomeini's return, he called on the revolutionaries to have 'patience'. ('Isn't that an oxymoron?' the revolutionaries sneered.) He filled his government with liberals who were keener on nationalism than political Islam. He put oil, the resource on which the economy depended, in the hands of men who suggested that Islam couldn't solve modern problems. Many of his ministers and bureaucrats were said to indulge in Western abominations, like the wearing of aftershave. Some of their wives walked about brazenly, with their hair uncovered. On the subject of the future Islamic Republic, they envisaged a tepid, Western-style democracy, scented with Islamic attar.

Such people couldn't be trusted to keep the country in the state of motion that was essential if the Revolution was to succeed. They couldn't be depended on to protect the Revolution's cardinal principle: the rejection of foreign ideology. Under them, the country could easily slide back into the US's sphere of influence. Bazargan and his friends might fudge the sacred duty of eliminating Israel. Their introspective, intellectual Islam was even more dangerous than secularism, because it assumed the garb of a friend. Bazargan was Iran's Kerensky. Like the liberal Kerensky, he would have to be destroyed.

The Imam started to undermine the provisional government. His supporters – clerics, influential traders, revolutionary activists – worked to bring about the clergy's supremacy. They sent their bullies to break up rallies staged by other groups: liberals, Kurdish nationalists, Marxists. Revolutionary committees were authorized to carry out arrests, executions and property confiscations.

Overseeing all this was the Imam's kitchen cabinet, the Islamic Revolutionary Council. The council was composed mostly of clerics who carried out or anticipated the Imam's wishes. They controlled revolutionary courts, which, independent of the justice ministry, handed out death sentences and prison terms to former officials from the Shah's dictatorship. They promulgated legislation by decree. They turned Bazargan into a knife without a blade.

I've seen a picture of Mr Zarif taken at this time: he looks supple, jackal-like, and his eyes are insouciant, and it's not the nihilistic insouciance of a Western boy, braving ideology – any ideology – to capture him. On the contrary: he has become pure ideology. God and Khomeini have let him into one of the most important secrets unveiled to humanity. Better still, he's taking part, furthering its interests. Mr Zarif is smiling in the photograph, deliriously happy to be alive.

It's after lunch. Persian after lunch starts after the nap that comes after the glass of black tea that comes after lunch. Mr Zarif won't go back to the office after this lunch. He'll go in tomorrow morning. He's taking off his socks, slapping them against nothing, against the air.

'You know, we saw everything from a revolutionary point of view, everything in revolutionary terms. I mean, if I said to someone: "Don't go home tonight, because we've got work to do," and they said, "Well actually we've got family coming round tonight and I really should be at home – perhaps another time . . ." Well, that would upset and shock me. I mean; what a strange set of priorities! Here we are, changing the world, and you want to go home and suck up to Aunt Maryam!'

He notices the socks in his hands. He goes over to the radiator and lays them on top. He's rolling up his sleeves. He disappears.

He's standing in front of the sink in the bathroom. He runs his right hand, soaking wet, down his face. He puts his wet fingers in his ears and wiggles them about. (Get out the wax.) He puts his wet fingers up his nostrils and wiggles them about. (Get out the snot.) He rubs his teeth with his fingers. He dribbles a little water over his widow's peak. He drags his wet right hand down his left forearm (from a point not higher than the elbow). He drags his left hand down his right forearm (from a point not higher than the elbow). He lifts up his legs, one after the other, and rubs the tops of both feet (right foot with right hand, and left foot with left hand).

He comes into the sitting room. He says, 'If I saw someone doing something suspicious, I'd immediately write a report on him, and if someone didn't have a beard I'd skip school and follow him. There was one guy in my street and I thought he was a leftist. Three Fridays in a row I followed him. Each time, a man with a beard rode by on a bicycle – the same man, each time. Naturally, I thought he'd been sent by God to help me in my investigation. And later I found out; no, he was a guy who lived in the neighbourhood, who happened to have a beard.'

He kneels.

At Mr Zarif's all-boys school, some of the female teachers believed that the Revolution had happened in the name of freedom – freedom of speech, thought, behaviour. (They had mistaken liberty – which means liberty from moral corruption and Godlessness – with amorality.) They took part in demonstrations that forced the Imam to back down on a decree that female civil servants cover their heads and wear shapeless clothes. There they were, persisting with their hip-hugging skirts and high-heeled boots.

The art teacher had cropped her hair, taking as an example one of the cops in Cagney and Lacey – Mr Zarif couldn't remember which. The Cagney and Lacey woman had favourites among the older boys. People whispered about what she got up to with her favourites.

(Mr Zarif stands, head bowed. He whispers: 'In the name of God the merciful and compassionate. Glory and thanksgiving be only to the God of the universe, who is merciful and compassionate and lord of the day of retribution. We worship none but you, and request help from none but you. Guide us along the right path, the path of those whom you have made secure, not the path of those who have lost their way. In the name of God, the merciful and compassionate, say that God is one. God needs nothing. He was not born, and did not procreate, and no one is like him.')

The ideologues were saying that the Revolution required several steps; the Shah's flight had been the first. Now, they said, it was the turn of the Communists and liberals and Westernized fun-lovers. There was a dangerous group, the Peoples' Mujahedin, which claimed to have reconciled Islam with Communism; the Prophet, they said, had been the first Marxist! (Later on, the Imam was to christen this group the Eclectics and, later still, the Hypocrites.) There were kids at school who daubed hammers and sickles on the playground wall. The head of the revolutionary committees said, 'We must purify society in order to renew it.' The question was: how?

One day, in a mosque that was known for its fervent and revolutionary congregation, Mr Zarif came across a group of people who had the answer. They were older than Mr Zarif – most of them were in their early twenties – and they called each other 'brother'. They wore trimmed beards and kept their shirts untucked. Even on hot days, they never rolled up their sleeves. One or two of them wore silver rings, with a star in the middle. Some of them had the piebald Palestinian scarf, the *kaffieh*, around their necks, and mentioned the Bekaa Valley in conversation. They grinned when Mr Zarif asked them whether they had spent time in Lebanon. Some of them seemed knowledgeable about automatic weapons and explosive devices.

(Leaning forward, hands on knees: 'The most elevated God is clean and pure.')

They were lovers. They loved the truth. They loved God and the Prophet. They loved the Imam and the clerics around him. They loved the Imam Hossein and the Imam Ali. More than anything, they loved their enemies – the liberals and Marxists, the Americans and the British agents. And the Zionists, of course. They would destroy them with their love.

They said they took orders from some clerics in Isfahan. (The clerics seemed to take their orders from people close to the Imam.) They were doing useful work: spreading propaganda, harassing opposition groups, encouraging citizens to denounce apologists for the former regime. Some of them were members of the Revolutionary Guard. Others were linked to the revolutionary committees. Some, Mr Zarif guessed, were members of an unofficial action group, called Hezbollah, though they were coy if asked.

(Kneeling over, forehead on a tablet of baked earth from Karbala: 'Great God is clean and pure.')

One by one, they and their allies were getting into the local bureaucracy. There was an increase in trimmed beards in the municipal corridors. There were more chadors. The Imam's supporters were making life difficult for civil servants who didn't say their prayers, or failed to turn up for indoctrination classes. The secularists had a choice: change your ways, and your appearance, or get out.

One Thursday evening, they let Mr Zarif join them in a small room next to the mosque. One of the younger lads picked up a microphone that was attached to an amp and started singing about Hossein's martyrdom. He had a fine voice. The others gathered in a tight circle, near the singer, and knelt inwards. In time with the lament, they brought their arms high above their heads, and down again, so that their hands thumped against their chests.

Gradually, the lament got faster. The arms rose and fell faster, like the pistons of a locomotive. Someone turned off the light and the men took off their shirts; their torsos glistened in the street light that came in from the window. Faster and faster, the lament went, until the singer's voice cracked; he started sobbing into the microphone. Inside the circle, the arms were rising and falling more swiftly; when the hands hit the chests, they made the sound of bones hitting hide. Drops of sweat fell off the end of Mr Zarif's nose. His arms ached. His chest felt raw.

Everyone was shouting: 'Hosseinhosseinhosseinhosseinhossein!' and hitting their chests as hard as they could.

After it was over, someone turned on the lights. Mr Zarif blinked. Everyone had red splotches on their chests. The room was humid. The lads put on their shirts. Then someone brought in tea and biscuits. Someone cracked a joke.

(Standing up, hands out in supplication: 'God! Favour us in this life and the next, and save us from the torment of hell.')

After they had tea, one of the men came over to Mr Zarif and introduced himself. He asked some questions, about Mr Zarif's political and religious convictions, and the situation at his school. Mr Zarif gave him what seemed – from the man's reactions – to be satisfactory answers. The man asked Mr Zarif to monitor the Communists, and the Mujahedin, at school. These groups had seized arms from armouries in the chaos that preceded and coincided with the Shah's flight. Their paymasters in Moscow were trying to take advantage of the situation, to suck Iran into their zone of influence.

(Sitting on his heels, hands on knees: 'In the name of God, on him be praise and glory. I bear witness that God is one and that Muhammad is his servant and Prophet. Greetings and the benediction of God on Muhammad and his followers.')

The following week, Mr Zarif and the other members of the gang followed the Communist kids. They found out where they lived, and discovered that their dads wore big moustaches, and called one another 'comrade'. Some of the dads worked at Isfahan's big iron works, which had Russian managers. One or two of them socialized with Russian families. The Russian families were poor and ugly.

One day, a couple of men arrived at the school to start political indoctrination. The men told the kids how to think about God and the Imam, and America and the Zionist Entity. When the principal saw that Mr Zarif was a friend of these men, he conceived for him a shaming fear. A kid of fifteen had become more powerful than he was.

Mr Zarif neglected his studies. He started doing sport, pumping iron, sticking out his chest. (He was growing a beard, though not fast enough for his liking). In school, he delivered harangues, handed round pamphlets. He organized prayer meetings in the playground. If he wanted to pass on a message to another boy, he would walk into the boy's class and whisper the message to him – the teacher would pretend not to notice.

Mr Zarif's boys got two of Cagney and Lacey's favourites into the school store, and asked them some questions. They learned that Cagney and Lacey was a closet Communist. Shortly after, quite a senior person from the Revolutionary Guard arrived at the school. He spent a long time in the principal's office. Cagney and Lacey was called in and invited to resign. The following day, at his word, ten of Mr Zarif's lads surrounded the Communists; there were bleeding noses. The hammers and sickles got fewer.

(Sitting on his heels: 'The peace and munificence of God be on Muhammad. Greetings on us and the right-acting servants of God. The peace and mercy and munificence of God be on you.')

A few months after the Revolution, the Communists planned a meeting that was to be addressed by a high-up Communist from Tehran. Thanks to a spy he had planted among them, Mr Zarif got wind of the meeting. He went early and got a good spot near the podium. Just as the speaker was being introduced, Mr Zarif ran onto the podium and landed a good one on his nose. Before anyone had time to react, he hurled himself into the section of the crowd that was thinnest. He was small enough, and fast enough, to get away with only a broken rib.

At the beginning of November 1979, radical students allied to the Imam seized the American Embassy in Tehran, taking the staff hostage. The students announced that they would release the hostages only when President Carter handed over the Shah, who had been allowed into America for cancer treatment. Bazargan resigned; his government had been trying to repair relations with the US. After Bazargan's departure, the Imam placed the government directly under the control of his kitchen cabinet, the Islamic Revolutionary Council.

Mr Zarif was delighted: he remembered that the Revolution was made up of steps.

❖ ❖ ❖

Nowadays, when people think of the mullahs' revenge, they think of Sadegh Khalkhali. There were scores of clerics who were more important than him; they actually took decisions, rather than implemented and interpreted the decisions of others, as Khalkhali did. Many of these mullahs were easier on the eye than Khalkhali; they had politer turns of phrase, more impressive qualifications. Khalkhali was a poor kid from the Azeri northwest, short on education outside the seminary, rotund, bald and coarse.

During the Shah's time, Khalkhali had upset people by writing a treatise depicting Cyrus the Great, founder of the Achaemenian empire and a figure whom the Shah admired, as a sodomite. He'd been imprisoned and internally exiled. Then, a few days after the Revolution, Khomeini appointed him to be a judge in the revolutionary court that was to try beneficiaries of the old regime and opponents of the new one. Khalkhali toured the country, trying monarchists and counter-revolutionaries. (Over a three-month period, he claimed to have condemned more than four hundred people to death. They included former senators, a radio presenter and a mob leader.) His pugnacious, fat face became as famous as his jokes, which often featured references to executions. V. S. Naipaul, who visited Khalkhali at the height of his notoriety, likened him to a jester at his own court.

Khalkhali made an indelible impression on Elaine Sciolino, an American journalist who witnessed one of the trials he presided over; she remembered him in a book she wrote two decades later. To counter the extremely hot weather, Sciolino recalls, Khalkhali removed his turban, cloak and socks, which must have made him look like a turnip. He sat on the floor and picked his toes while hearing the evidence against a defendant. He repeatedly left the room during the testimony of witnesses.

His most famous victim was Amir Abbas Hoveida, and Khalkhali must have enjoyed that bit of business. Hoveida was Khalkhali's antithesis, thirteen years the Shah's prime minister, a man whose Northampton brogues Khalkhali could not, before the Revolution, have dreamed of polishing. Hoveida was a francophone, but he also knew Arabic – the Arabic of Beirut society, not the Qoran. Even after their divorce, Hoveida's wife made sure that a fresh orchid reached him every morning for his buttonhole. He'd not been personally venal or murderous, but he'd closed his eyes to the atrocities of others. Khalkhali charged him with waging war on God and corruption on earth. Over two court sessions, separated by several weeks, Khalkhali pounded the defendant's moral ambivalence like saffron under a pestle.

At lunchtime on Hoveida's last day, Khalkhali reports in his memoir, the prisoner was treated to a repast of rice, lamb and broad beans. Khalkhali claims to have made do with bread and cheese. (Next to photographs of him, excessively crapulent, this ascetic self-portrayal is unconvincing.) During the afternoon session, Khalkhali didn't allow Hoveida a defence counsel, nor was a jury present. As the presiding judge, Khalkhali didn't pretend to be impartial; in the vehemence of his harangues, he rivalled the prosecutor.

By trying Hoveida, Khalkhali jabbed his finger in Bazargan's eye. Bazargan disapproved of the revolutionary court – he was planning for Hoveida an exemplary trial that would establish the Revolution's reputation for justice and moderation. But Khalkhali, who plausibly claims to have taken hints from Khomeini, had different ideas. He gave orders that no one was to be allowed out of the prison where the trial was taking place. To ensure that word didn't reach Bazargan, he locked the prison telephones in a fridge. And so Hoveida was sentenced and shot in the prison courtyard. His final words were patrician, and a bit surprised: 'It wasn't meant to end like this.'

Khalkhali's theatre travelled on. It gave perhaps its most memorable performance at a famous shrine in south Tehran. Khalkhali and two hundred revolutionary militiamen set out to destroy the Pahlavi family vault, which was in the shrine's precincts. Khalkhali was opposed by the government and by the resilience of the granite structure. The spades and picks used by the Revolutionary Guard proved insufficient. Khalkhali called for reinforcements. (National television was already on site to record his endeavour.) Bulldozers and cranes arrived, but the tomb withstood. At ten o'clock that night, the valiant revolutionaries went home to bed.

In his memoir, Khalkhali craves his readers' indulgence: 'Perhaps you don't grasp how strong they'd made this tomb.' But he was not deterred; the tomb would have to be blown up, by degrees. And when, after twenty epic days, the job was done, and the dust of imperial bones blended with the smell of cordite, 'the sound of cheers and joy rose from the people, and the enthusiasm and joy were indescribable'.

You can't make an omelette without breaking eggs. That might be Khalkhali's epitaph.

❖   ❖   ❖

Today, Mr and Mrs Zarif are coming to lunch with us, here in Elahiyeh. I wonder what they will make of Bita, my wife, and what she'll make of them.

Elahiyeh is a desirable suburb on the slopes of north Tehran. It used to be so green that, even in midsummer, you had to sleep with a light blanket. The British and Russian embassies kept grand legations in Elahiyeh, to which their respective ambassadors decamped in the spring. Now, the compounds remain but most of the gardens have been built over. Elahiyeh is rarely more than two or three degrees cooler than the dustbowl of south Tehran.

Elahiyeh's name is derived from the name of God in Arabic, Allah, but few places in Iran are more reputed for impiety. Behind entry gates crowned with barbed wire, illicit booze is consumed and dancing committed by mixed assemblies. Anecdotal evidence suggests that sex happens between men and women who aren't married to each other. The Islamic Republic is an avoidable botheration.

In the Shah's time, the area was inhabited by suave monarchists who built Swiss-style chalets. Fearing for their liberty after the Revolution – many of them had taken part in the Shah's oppression, or dipped into the public purse – they fled. The new regime appropriated their houses and grounds, building on them or turning them into, say, a sports club, the families of a privileged caste of civil servants.

Elahiyeh's present inhabitants are an uncouth upper class. They have done well in recent years out of high oil prices. They inhabit marble-clad apartments in escapist blocks and enjoy the view during the rare instances when smog hasn't settled in the lap of the Alborz Mountains. Many of them have residence rights and property abroad – the Revolution taught them that it pays to keep your options open.

It's difficult to ascertain exactly where their money comes from. Knowing the right people has a lot to do with it. They are terrific name-droppers. Having access to commodities beyond the reach of the common man – foreign currency at preferential rates, import licences – is also important. Their skill is acquiring what exists in artificially small quantities and selling it at a price reflective of this scarcity. Their wives take lovers and visit a French-educated psychologist downtown.

Their teenage daughters, matchsticks marinated in Chanel, are yanking up their coats; in recent years, hems have drifted above the knee for the first time since the Revolution. Their favourite activities are having nose jobs – there is one model: *retroussé* – buying illegally imported Italian shoes and rearranging their headscarves in public, by mistake on purpose exhibiting their hair.

The daughters gather on a Thursday night, outside pizza parlours and coffee shops, discharging arch glances and pollinating scents. They're treading water while their parents find them a mate. (Likely as not, he will be their first cousin – the families know each other, and the *mehriyeh*, a kind of pre-nup, will not be prohibitive.)

They are courted, if the word is applicable, by boys who wear a minimalist variant on the goatee, driving Pop's sedan. A chance meeting in a coffee shop; a telephone number flung into a passing car – such are the first moves. Oral sex is, of necessity, popular; there will be a great to-do if the girl doesn't bloody her wedding bed. In case of penetration, however, all is not lost. A discreet doctor can usually be found to sew up the offending hymen.

There's a hollow thrill to be got from bettering the morals police. (They cruise Elahiyeh in their Land Cruisers, looking for miscreants to shake down for a few dollars, smelling breath for alcohol, rummaging through handbags for condoms.) For the rich kids, it's the best way of getting back at the state, at parents, at the predictability of life.

In a strange way, Elahiyeh's social vacuum suits us, too. We like the traditional notion of an Iranian community, but are not sure we could inhabit one. Unlike almost everywhere else, you can live in Elahiyeh as you can in a Western city: in peace and anonymity.

Before 1979, Bita's parents had nice ministry positions; both regarded a deputy ministership or another senior bureaucratic post as their due. Bita and her younger brother – a second brother was born on the eve of the Revolution – led blameless, privileged lives.

There were three choices when it came to educating your children: the French school, the German school and the American school. (You didn't send your child willingly to an Iranian school; foreign languages and contacts were indispensable aids to getting on in the world.) The trouble with the American school was that its graduates spoke Persian with an American accent. There was no German connection in Bita's family. Her mother, on the other hand, had studied law in Paris, so Bita was sent to the French school. It was run by nuns. Each year, on the anniversary of her martyrdom, the school commemorated the exemplary life of Joan of Arc.

Bita wore a dark-blue collarless tutu over a white T-shirt. In winter, she wore a roll neck jumper over the T-shirt. If the driver was late collecting her after school, she would wander down nearby Lalehzar, Tehran's Pigalle, where there were whores and the smell of alcohol, and ornate cinemas with *putti* on the ceilings. In the summer evenings, when her parents were out, she would go swimming in pools that belonged to the parents of her friends. She and her friends danced to Googoosh, Iran's answer to Shirley Bassey.

They admired Farah, the Shah's third wife. It's arguable that Farah was not as exquisite as wife number two, an Isfahani whom the Shah abandoned for failing to sire. But, she was tall, wore fabulous clothes and had an artistic eye. She was an alumna of the French school and came to visit.

In 1978, there were riots and atrocities. Bita got used to the sounds of firing and being sent home early from school – and the worried look on the face of Ma Soeur Louise. She didn't realize that she and her friends, and Farah and the Shah and the whores of Lalehzar, were the reason for the hatred.

And so the Shah left. An old man with frightening eyes came. The French school was closed. (Of course it was; it was named after a Roman Catholic saint!) A lot of the girls, including Bita, were removed to an Iranian school where French was taught. Friends started leaving. First, the foreigners and the Jews, and the Bahais – members of a religious sect, originally an offshoot of Islam, that had been favoured by the Shah. One day, little Ziba would come to school. The next, she'd be gone. A few weeks later, her family would surface in Orange County, California.

It seemed to Bita that everything had been turned upside down. The people who were now giving orders looked like the people who had taken orders before. In the past, her mother and father had been on top. Now, they were at the bottom. If they wanted to get something done, they had to flatter coarse men with beards and rosaries. In the past, Bita had associated beards with building workers and dervishes. Now, everyone was growing them; you had to, if you wanted to get on.

A few months into the Revolution, Bita's new school was closed and she went to another. They didn't teach French at the new school. Arabic, the language of the Holy Qoran, was compulsory. The girls had to wear headscarves and long coats. They were told to despise the wearing of ribbons in hair, and bare ankles. In the streets, there were Hezbollahis patrolling, checking peoples' adherence to Islamic rules concerning dress and behaviour. They threw acid in the faces of women who were inappropriately made up.

Bita had lived for colour. It was as important to her as the sun. The Revolution had killed colour, declared it to be evil.

◆ ◆ ◆

Mr Zarif had delivered his school to the Revolution; in the precincts, he was unchallengeable. He turned his attention to a Qoranic injunction that Muslims promote virtue and prevent vice. It meant implementing Islamic law and practices, eradicating decadent ways of behaving. It meant starting at the bottom of society. He and the gang started hanging around parks and shopping centres. They would approach boys who were chatting to girls and ask, 'What is your relationship? Is this woman your sister? Why are you talking to her?' If they got an unsatisfactory answer, they'd hustle the boy away and tear off a few shirt buttons. They'd tell the girl: 'Bleached jeans are a sign of American cultural corruption. Go home and put on Islamic clothes.'

The ban on booze was hitting the alcoholics. Liquor prices had rocketed. Every morning, a park or a vacant lot yielded up a new body, full of petrol, turpentine, meths – anything they could get their hands on. Mr Zarif felt that society was being cleansed, spewing harmful matter. He was learning Arabic, the language of the Holy Qoran.

Sometimes, he and his lads caught boys and girls flirting in shops, under the cover of deciding on a purchase. Mr Zarif and the gang would smash the windows of shops where such things went on and spoil some of the merchandise. If they saw girls flouncing in a park, they seized their handbags and tipped out the contents. 'Who do you wear make-up for?' they demanded. 'What is that music cassette you've bought? Haven't you heard what the Imam said about Western culture?' If they came across a young man wearing a Led Zeppelin T-shirt, they said: 'Your hair is longer than Islam permits. Everyone should groom himself as the Prophet did. Here; let us cut it for you.'

They would deliver serious offenders to the boys at the mosque. The boys would consult one of the mullahs and get a sentence passed. Whippings would be administered, in accordance with Islamic law. The gang's effectiveness was enhanced by the recruitment of two middle-aged women with long nails; they seemed to enjoy scratching the faces of pretty girls who were resistant to the Islamic dress code.

❖ ❖ ❖

The doorbell rings. It's the Zarifs. We've cooked Indian food, because we reckon that Mr and Mrs Zarif should be open to new experiences.

Not too new. Bita is wearing her headscarf. She's careful not to put out her hand to shake Mr Zarif's. She helps Mrs Zarif get out of her black chador for outdoors, and into her colourful indoor chador. Mr and Mrs Zarif look around for indoor slippers to put on. But we don't ask people to take off their shoes when they enter our house. There aren't any slippers available. Mr and Mrs Zarif take off their shoes and walk on in their socks.

'What a house!' they both say it at the same time. They look at Bita. (She's the interior designer.)

The hall is burgundy. (My father-in-law says it looks like a nightclub.)

There is a batik wall hanging depicting the Hindu goddess Durga, wearing a necklace of human skulls.

The sitting room is two shades of tangerine. There's a picture of a woman in a bright red dress and a challenging stare, standing next to an androgyne with diaphanous blue skin and yellow hair. There are red-backed chairs and an Indian sari turned into curtains, and a dark green sofa from the 1940s, and a green tribal tunic with red paisley lining put in a frame and attached to the wall. The bolsters are richly coloured and patterned. There are riotous Baktiari carpets, Armenian rugs.

Mr Zarif is wearing a grey shirt, and grey trousers, and white socks. His house has white walls.

As we sit down to eat, I wonder whether he ever threw acid in the face of a girl who had red on her lips, or hair escaping from her headscarf.

CHAPTER THREE
A Sacred Calling

One morning in the autumn I found myself in the back seat of a stationary taxi, facing due south, inhaling exhaust fumes. The authorities call this road an autobahn, because it's meant to be quick and efficient. They have flanked it with lush verges on which they squander the city's meagre water resources. I don't think the former mayor, Ghollam-Hossein Karbaschi, who built this and most of Tehran's other freeways, listened to foreign experts when he was drawing up his ideas on public transport. Had he done so, he would have learned that more asphalt does not lead to less traffic, but to more. Karbaschi's urban arteries do not race. They loop clownishly. During the rush hour they atrophy.

On the car radio, a woman greeted us. 'To all you respected drivers and dear, dear bureaucrats, to you conscientious teachers and workmen, I say: *Salaam* and good morning! To all the beloved professors and students of the Islamic world, I say: Good morning!'

According to the scientists, we in Tehran take in seven and a half times the amount of carbon monoxide that is considered safe. This information starts to mean something only after ten days or two weeks without rain, without wind. One morning, you look towards the Alborz Mountains and they're not there. Rather, they're impressionistically there. They're lurking behind a haze that's pink-grey, like the gills of an old fish. If you go out for long, you get cruel headaches for which lemon juice and olives are the recommended cures. Windless weekdays are said to carry away scores of old people, all of them poisoned. In the town centre, there's a pollution meter whose optimistic readings, naturally, no one believes. The sunsets look like nuclear winters.

The woman speaking on the radio sounded as if she was on LSD. She said: 'I think it would be a good idea for us to perform some simple acts that enable us to start the day in fine fettle. If the window of the car you're in is closed against the cool of the morning, start by asking the driver if he would mind winding it down. Actually, why don't I ask him myself? "Mr Driver? Would you mind lowering your window a little?" And to all those housewives at home, I say: open the window a bit, the weather's splendid!'

Tehran has too many cars and not enough buses. There's a plan to replace fifteen thousand elderly taxis. There's a plan to give out loans so that taxi drivers can run their vehicles on compressed natural gas. There's a plan to extend the metro, which at present has limited reach and is overwhelmed by the rush hour. There's a plan to increase public awareness, to tell the middle class it's not below their dignity to use public transport. Plans, plans.

'Take a deep breath, and keep it a few seconds inside your chest. Now, slowly let it out again. Exactly! During the next song, I want you to do this several times.'

There should be a plan to teach Iranians how to drive. On the road, there's no law, no *ta'aruf*. There's no inside or outside or middle lane; the heavier the traffic, the more lanes come spontaneously into being, and the narrower they are. There's no indicating left or right. There are pedestrians who can't be bothered to take the pedestrian bridges, crossing the motorway like morons. Some evenings, when the kids are out, with the ducking and weaving at extraordinary speeds, you might think you're in a rally or a computer game. Or you could think of it this way: the vehicle you're in is a laggard sperm and the end of the freeway is the last egg available to humanity.

I've seen cars prostrate over advertising hoardings; I've seen a compressed pedestrian dead like a slug in the middle of the road. I've seen cars skittle mopeds – no helmets of course, that would be sissy – and drive on regardless. Drivers communicate by leaning on their horns and flashing their headlights. They use symbols: the thumbs-up (a rough equivalent of the finger), the clenched fist (a bit worse). Tempers fray. Once, as a passenger in a taxi, I found myself leaning out of the window and deploying a Turkish profanity that I had learned while living in Ankara but had never, on account of its considerable obsceneness, dared to use.

The elderly taxis are Paykans. In winter, Paykan drivers stick a piece of cardboard across the grille, giving the car the appearance of an asthmatic with a hanky in front of his mouth. Paykan means arrow, but the Paykan is as unerring as the Hillman Hunter, its almost identical antecedent from the 1960s, was sharp-nosed and predatory. In the old days, Paykans were mainly British-made and assembled in Iran. But the British don't make Paykan parts any more, and 97 per cent of every Paykan is Iranian. I have been told that every new Paykan rolls off the production line with an average of two hundred faults. This is the reason why a fifteen-year-old Paykan, which has more British parts, will cost you more than a new one.

'And now it's the turn of the smile. Everyone smile to everyone! The rose of a smile will beautify your face. The scientists have established that people who smile in response to daily challenges are more likely to retain their health. Don't frown!'

Something happened and we started to move. Sometimes, it's not obvious why these traffic jams happen, and why they stop. It's one of the mysteries of Tehran.

❈ ❈ ❈

In the 1990s, Karbaschi let the magnates into north Tehran, where they developed Elahiyeh and other neighbourhoods with little regard for taste or safety. (It's not unknown for new buildings to subside as a result of vibrations from nearby building sites.) The city's infrastructure couldn't keep up with the pace of growth, and there was a bad smell of impropriety. When Karbaschi was jailed in 1998, everyone knew his trial was politically motivated. But no one suggested that his municipal empire wasn't corrupt.

Now, four years after he was pardoned and freed, Karbaschi is infrequently criticized. His freeways, his skyline, his parks and his cultural centres: they symbolized a regeneration, Tehran's version of the building boom that bulldozed and revived Europe's cities in the 1950s. Karbaschi was announcing: the War's over. Let us look to the future.

But a revolutionary state can't look to the future. The Revolution is everything, and it has already happened. The War was the Revolution's crescendo, so the authorities have preserved it. Living in Tehran is like listening to the sea in a shell.

The authorities made the War part of the fabric. They put it on the city maps. As casualty figures rose, so the localities started changing. Thousands of streets called after nightingales, angels and pomegranates were given new names. Martyr Akbar Sherafat (this was the street where he grew up; his parents still occupy a flat in number sixty-one); Martyr Soufian (his daughter was born a few days after an Iraqi shell scattered bits of him over the front); the Martyrs Mohsenian – two brothers whose faces, smiling down from heaven, have been painted on a wall.

In the process of finding a friend's house, you commemorate heroes: 'Excuse me, madam, where's Martyr Khoshbakht Alley?'

'Well, you go down Martyr Abbasian Street, turn right into Martyr Araki Street, and then turn left immediately after the Martyr Paki General Hospital . . .'

So much for the little men with their little places; the prestige memorials – the boulevards and autobahns – are reserved for the dead elite. In the north of Tehran, there's Sadr Autobahn – that's Iraq's Ayatollah al-Sadr, Iraqi Shi'ite, whom Saddam Hussein executed for sedition. Sadr is tributary to the main north–south autobahn, Modarres (Ayatollah Modarres, who was known for his opposition to the last Shah but one). Closer to the Square of the Seventh of Tir, there's Beheshti Avenue. (Ayatollah Beheshti was the Islamic Republic's first chief justice.) Before the Revolution, Beheshti Street was called Abbasabad.

My taxi was going on slowly. I saw that scaffolding was up in front of a mural that had interested me since my arrival in Iran. Men in overalls were sitting on the scaffolding, under a canopy. There were pots that I assumed to be full of paint; they were preparing to paint over the mural.

The mural showed a dead man, a martyr, lying in his bier, with his daughter standing over him, holding a rose. The daughter couldn't have been more than four years old, but she wasn't looking down on her father with the exuberant grief that you might expect. Her expression said: 'I understand. You were my father but, more important, you were a Muslim. Having weighed your competing responsibilities, you went off to defend the Revolution, and Islam, from the Iraqi rapists. Good for you.'

I couldn't imagine the little girl giggling, or whining, or tugging at her mother's chador and demanding ice cream. Her dress was fanatically Islamic; who ever heard of a four-year-old wearing a black smock to cover her hair, and a chador over that, with not so much as a lock on display? A four-year-old alive to the diabolical temptation represented by a woman's hair? She wasn't a girl, but an idea.

We passed Mottahari Street (former name: Peacock Throne Street) – that's Ayatollah Mottahari, Khomeini's colleague and friend, who was assassinated a few months after the Revolution. We reached the Square of the Seventh of Tir – former name: the Square of the Twenty-Fifth of Shahrivar, the date of the Shah's accession to the throne. Not a square in the Western sense, or a grassy *maidan* in the Indian – more an oxbow for Karbaschi's meandering freeway, with a scum of shared taxis and cars and buses.

On the Seventh of Tir 1360 – that's the Iranian calendar date for 21 June 1981 – a huge explosion that is thought to have been planted by the Hypocrites killed seventy-two people, including Beheshti, four cabinet ministers and other bigwigs. (Two more later died of their wounds.) On a wall overlooking the square there is a mural of Beheshti with his wiry beard and olive-stone eyes. Underneath, there is his eccentric adumbration of Iran's foreign policy: 'Let America be irritated by us; let it be so irritated, it dies.'

The carnage of the Seventh of Tir convinced Khomeini that there could be no mercy. The enemy, the Communists, liberals and pseudo-Islamists, had to be destroyed. In the months that followed, thousands of members and sympathizers of the Mujahedin and other opposition groups were executed. On 18 and 19 September 1981: 182 (according to official figures). On 27 September 1981: 153.

We entered Roosevelt – it acquired a new name after the Revolution, but everyone still calls it Roosevelt. We passed the Nest of Spies. It's the regime's name for the former US Embassy. Low-slung walls: easy enough for the students to get over. I remembered pictures from *Time* magazine at the end of 1979, of the hostage-takers using an American flag to carry away rubbish from the embassy compound, and a lurid Khomeini, Hammer Horror with blood-red irises, on the cover.

A few months before I'd visited a temporary exhibition at the Nest of Spies. The people had come to smell America. They'd come to look at the eavesdropping equipment that the embassy staff had used, and the shredders and incinerators they'd fed with documents as the students took over the embassy. (The students then spent months piecing together the shredded material. Some of this, they were able to claim, implicated their domestic rivals in CIA plotting. This was helpful to Khomeini, who used the findings to discredit his opponents.)

The organizers of the exhibition had placed dummies of American diplomats around a table, in a soundproof room that had apparently been used for secret meetings. As a visitor to the exhibition, you stood outside the room, which was made of two thick panes of glass with a vacuum between them, and looked in at the Americans. They wore ties: a Western affectation. They were seated on chairs: a kind of enthronement. They had crossed their legs, or splayed them, showing off immodest American crotches: canine. As you stood there, pressed up against the glass, and viewed their washed-out complexions and ugly auburn hair, you could imagine them talking over ways to control Iran, to defeat Islam. At the end of the working day, you could imagine them drinking beer and taking a slut for the night. That was what Americans did, wasn't it?

We carried on south. We crossed a flyover. On one side, the houses had not been fully demolished – just enough to allow the flyover to be built. They were half-houses. The upstairs rooms still had wallpaper. The grid of south Tehran started to take shape. Scraps of yellow and turquoise tile were visible on the older façades, and rust-coloured roofs. There was less building activity in this part of the town and more traffic. The women mostly wore chadors. A different town, conservative and claustrophobic.

Sometimes, I've wondered what it would be like to live here. There would be a mode of conduct, proximity to the neighbours, a feeling of impermanence. These old communities are under attack – by unemployment and highly adulterated heroin at fifty cents a hit, by women who aren't family and the influx of migrants from the provinces. Nothing stays the same. A neighbour leaving, another taking his place, a divorce, a business success, an iron ball crashing into a corner shop.

The defences are religion and the watchful eyes of neighbours, the chador and Islam. If the community is an island, and if the roads and bazaars full of strangers are the sea around them, then people behave themselves on the island and swim free in the fathomless waters of moral decay.

Then, we were caught in the bazaar traffic. Small vans carrying carpets and cans and wooden palettes on their sides. Men pushing carts: the porters, the lowest form of bazaar life. The day before, the bazaar had closed its doors in protest at an aggressive speech made by President Bush. It was to show America that Iranians were united in their continued hatred for the Great Satan.

As we approached the South Terminal, I looked out for a large black building, a plant that produced vegetable oil, which I was used to seeing at the roadside. But the factory was doubled over – in pain, badly winded. The roof had collapsed. One of the chimneys had toppled.

I got out of the taxi, holding my bag, and turned to face the Peugeot drivers.

'Isfahaaaan! Isfahaaaan!'

One of them came up to me. He had a bronze complexion, purplish lips. 'Isfahan! Leaving right now!' His face was convulsed by the opiate's bonhomie. (In Iran, the masses have both religion and opium.) His hand gripped me insolently.

I picked another driver, one with a clean moustache and an ironed shirt. The back seat of his Peugeot was occupied by a man in his twenties and another chap with a beard. I took the third place. A young couple shared the front passenger seat and fed each other crisps.

We moved off. The driver shifted position in his seat, hunched over the wheel. He flicked the gears with his palms and ran his hands through his shiny hair. There was a short conversation about what music we would listen to. The field was narrowed down to the titans of Turkish pop: Tarkan or Ibrahim. Ibrahim won. The driver pushed Ibrahim into the cassette player with the tips of his fingers. He lit his cigarette, but not before putting it in a mahogany-coloured holder. Every elegant move seemed designed to beguile the senseless boredom of his hours. We left south Tehran.

The sun in my eyes; Ibrahim lamenting through his moustache; the proximity of the five others; cigarettes; the speed and a rococo driver.

I thought: why don't I have a car? Now that baby's on the way, we'll have to get a car. Must be air-conditioned. But expensive! Government monopoly over car making, and demand far exceeds output: prices artificially high. Paykan? Forget it; Bita would sooner walk. Best alternative? Eight grand for a Kia Pride, a Korean-designed paper cup set on Smarties.

I was feeling sick and we were pelting along. We were driving through Zahra's Heaven, the main cemetery in south Tehran. Seventy thousand dead soldiers in there. Other fathers' sons, other men's exercise, mirth, matter.

Then we were speeding down dust tracks that had been thrown across fields of barley. We could follow the asphalt, but that would take us through the tollbooths at the beginning of the motorway. This way, we'd emerge onto the motorway a few kilometres beyond the tollbooths and cruise for free.

We skidded onto the motorway. One hundred and fifty kilometres an hour, in an Iranian-built GLX 2000. Tired driver, straight road; he could fall asleep at any moment. One careless bolt, cruelly loosening. That's all it would take. I looked at the other passengers. The bearded chap was silently mouthing an invocation, again and again, using dead time to accumulate credit with God. The couple had fallen asleep entwined. No one was thinking about seat belts. If we had to brake suddenly, we'd be scattered over the tarmac.

MR DRIVER, HAVE YOU CONSIDERED THAT EVERY ACTION HAS A CONSEQUENCE? WHAT DO YOU THINK OF CAUSE AND EFFECT?

The thought of never seeing my wife again. Or the little one, when he/she emerged. Right now just walnut size or strawberry size or whatever. A thing, not a person, but promising. Something I will love, and will love me, even if I prove to be unworthy.

MR DRIVER, WHY ARE YOU DRIVING SO FAST?

I tapped the driver on the shoulder.

'Mr Driver?'

He looked at me in the rear-view. He turned down the music a little bit and said: 'You don't like Ibrahim?' The young man was looking at me.

'No, no, Ibrahim's fine, I was wondering, could you drive at a more . . . er' – I groped for the word – '*reasonable* speed?'

The driver's expression in the rear-view mirror was puzzled. What did 'reasonable' mean? What did I want him to do?

He put his foot down. The speedometer gave up the ghost.

❈ ❈ ❈

I was in Isfahan, zigzagging towards the Shah's Square. (New name: the Imam's Square.) I was on my way to meet a cleric called Mr Rafi'i, to talk about the War. Bobbing above the surrounding houses was the blue dome of the Mosque of the Shah (new name: Mosque of the Imam), which dominates the southern end of the square. As I walked I passed iron gates that led into new tenements, or into an old courtyard that may have contained a fig tree and a tethered goat. The tight turning streets were still and baking, and my mouth was dry. I wanted to be close to the mosque, with its shadows and ablutions pool, and its moist revetments.

The normal way into the Shah's Square is through the roads and lanes that feed it from east and west, or from the bazaar, which debouches into it from the north. But my hotel was south of the square and I didn't feel like walking half its length before entering it from the side. I was trying a short cut. Having approached the mosque from behind, I would surely come across a passage or lane that ran alongside it, and that would take me into the square. I pursued the dome.

After a few minutes, I rounded a bend and met a massive brick wall. Lying in the dust, there were bits of broken tile – yellow and turquoise and blue. I realized I was under the tiled dome; it had moulted faience. I was standing at the foot of the rear wall of the main dome chamber.

If you approach the east end of a Gothic cathedral, you'll come across the apse's satisfying bulge, some gargoyles, a ribcage of flying buttresses. The Ottoman mosques are mystic spheres; whatever your viewpoint, there is always a painstaking accretion – of domes and half-domes, ascending to the main dome, and thence to heaven. Both have been conceived sculpturally. You're allowed to approach from all directions. But here: this rude wall!

When I stood a little to one side of the wall, I could see much of the mosque's skyline. From the Shah's Square: a pageant. From this side: a chaos of features and perspectives, without colour. I made out the western vaulted portico, or *aivan*. Viewed from the mosque courtyard, it is dazzling; the lavish stalactite decoration is intensified by mosaics and tiles. Now, from behind, it was unkempt, pregnant with its own vault, made of old bricks.

I had always assumed that the upstairs bays over the small vaulted shop fronts that flanked the mosque were the façades for storerooms and cells. Viewing them from the rear, I realized they were a screen. The Shah's Square was a theatre and I had blundered backstage.

There was no corridor past the mosque. I retraced my steps and walked north along the main road running parallel to the square. I turned right and felt the anticipation the architects intended I should feel. I entered the square where they plotted I should enter and saw what they wanted me to see. A vast bounded esplanade, bay upon bay, greatly monotonous. At the northern end: the entrance to the bazaar. At the southern end, the Mosque of the Shah. About two-thirds down, opposite one another: the Palace of Ali Qapu, and the Mosque of Sheikh Lutfollah.

The portico of Ali Qapu was crowned by a veranda with a roof supported by spindly wooden legs. From here, Abbas had watched polo matches, executions and military parades that took place in his honour. Beyond, at the southern end of the square, shivered the Mosque of the Shah. Slender minarets crowning the entrance portal; the dome's colossal bulk and the harmonious disposition of traditional forms – four *aivans*, facing each other around a courtyard. With one renowned aberration: in order to face Mecca, the entire mosque after the entrance portal had been oriented obliquely.

In Iran, the beloved monuments are not buildings but gardens. The most respected engineer does not make roads but the underwater channels that carry the water that cools the houses and moistens the desert. The great mosques are clay cups, and the Mosque of the Shah has water to the brim.

Up close, the tiles are coarse. Their prodigious acreage is almost unattractive. From a distance, however, you long to be submerged.

Around the square, families were claiming the garden and pavements that had been laid over Abbas's esplanade. There were picnics on the grass and girls playing badminton. Mothers chewed sunflower seeds and spat out the shells, while their husbands lit paraffin stoves. Urchin boys clung to the axle-bars of phaetons that propelled gently the well-to-do. Every now and then a shuttlecock would rise and fall before the dome of the little Mosque of Sheikh Lutfollah, like a tropical bird in front of a tapestry.

The Sheikh Lutfollah is one of the triumphs of all architecture. It has no courtyard, no minaret. The dome is low and made of pink, washed bricks, articulated by a broad, spreading rose tree inlaid in black and white. This dome catches the light shyly; the inlay is glazed, but not the bricks. The dome floats upon an *aivan* of typical ostentation – but askew, for the chamber has been placed twenty metres to the north. Why?

I crossed the square and went in. A small corridor opened off to the left: dark and dimly gleaming. I followed the corridor, and the darkness virtually obscured the tiles on the walls and vault. A few paces on, I was forced to take another turn, to the right, thick-wrapped in the corridor.

Ahead, a shaft of light, strained through window tracery, appeared from a wall two metres in girth. (The walls need muscle, to withstand the dome's thrust.) The shaft of light pointed like a Caravaggio. I followed it, turning right and standing at the entrance to the dome chamber. The dark corridors had disoriented me and made me forget where I was in relation to the square outside. I'd taken no more than twenty-five paces.

A man and a woman and a little girl were standing under the dome, talking. There was one other person in the sanctuary, a heavy man.

The architect – whose name, Muhammad Reza b. Hossein, is inscribed in the sanctuary – skewed the dome chamber so it faces Mecca, but that is the extent of the Lutfollah's resemblance to the Mosque of the Shah. In the Lutfollah, the dome chamber's orientation is not an ostentatious oddity, but hidden, subordinated to the serenity of the whole.

The light in the sanctuary was more plentiful but dappled through the tracery of windows in the drum and by the glazed and unglazed surfaces around the chamber. It illuminated, seemingly at random, a section of the inscription bands and a bit of ochre wall inlaid with arabesques, and a clenched turquoise knuckle, part of a frame for one of the arches.

Imagine Abbas, at prayer in his oratory, his head bared and vulnerable, fluid sunlight catching his shoulder.

The little girl, idling while her parents examined the enamelled lectern, gazed up at the dome, put her arms out, and whirled.

The thickset man addressed me: 'Mr Duplex.'

I said, 'Mr Rafi'i?'

The man said: 'Can you smell him?'

I sniffed.

He tried again: 'Can you smell God?'

❖ ❖ ❖

I followed Mr Rafi'i back along the corridor, towards the mosque entrance. He stopped outside a door that I hadn't noticed and pushed it open. We looked in on a plain cell. 'The Sheikh was Abbas the Great's father-in-law,' he said. 'This is where he prayed, and where he was buried.' Mr Rafi'i seemed to approve of the Sheikh's simple tastes.

We entered the square, and I looked at Mr Rafi'i. In the half-light of the mosque, my attention had been drawn by his thick torso and neck – not a taut musculature but a ragged peasant virility. His face was red, bulging. He wore a check shirt and dusty baggy black trousers.

'I just got in from my fields,' he said, guiding me across the square. 'Next week, we're going to start harvesting. But it's a busy time of year; I have to teach at the same time.'

Mr Rafi'i's subject was the sayings, sermons and letters of the Imam Ali. For Shi'as, Ali is the supreme example of a just and generous sovereign. During his caliphate, he is said to have bought two shirts and offered the finer of the two to his servant. His judges were so independent, one found against him in a case.

As my ears got used to Mr Rafi'i's rural accent, my eyes were drawn to his forehead. Many Shi'as have a purplish blotch there, from the baked tablet of earth they press down upon as they pray. Mr Rafi'i's blotch had gained a crust, with small features of its own. It seemed to laugh whenever he did – a wizened sprite, living in his head.

'Here's my horse,' said Mr Rafi'i, pointing to an old motorbike with a hempen packsaddle that might have been designed for a donkey. 'Get on.'

We went hoarsely down the little streets, into a main road that carried us, by way of one of the newer bridges, across the river. We strained up the hill on the other side, towards a shelf of mountains. We turned well before the mountains, continued for a couple of hundred metres and stopped at the gate of the Rose Garden of the Martyrs. We dismounted and Mr Rafi'i mouthed a greeting to the martyrs.

There are some seven thousand of them and each grave is surmounted by a metal frame that contains a photograph of the man in the grave. The graves are bunched, like copses, one copse for each major engagement. They represent a fraction of the martyrs from the province of Isfahan – I've heard of villages with a population of two or three hundred, and a score of graves in the War cemetery. The martyr's families would come each week, Mr Rafi'i explained, usually on Thursday evenings. He said: 'Come and meet my friends.'

He'd been their ally, their chaplain. He remembered the occupants of many graves; there was barely one whose name meant nothing to him. The photographs were formal, taken in a studio to commemorate earthly achievement – a school diploma, an engagement to be married. Perhaps a mother had sensed the coming martyrdom and requested a memorial pose.

'These boys were nothing like the boys you see on the streets today. Nothing! They were clean! And they were fighting for God. They were fighting for the government of Ali; they longed for his caliphate.'

He pointed. The photograph depicted a very young boy with the beginnings of a beard. 'He and his cousin died on the same day, coming back across the marsh. He was a good boy. They didn't find his body.'

'His grave's empty?'

'No, no . . . Listen! His parents tracked down some survivors from the operation, and got conflicting reports. One said he'd seen their boy being crushed by a tank. Another said he was electrocuted when the enemy diverted power into the marsh . . .'

Mr Rafi'i paused; he'd seen someone he knew, a bald man holding a watering can over a grave.

Mr Rafi'i called out: *'Salaam Aleikum!'* The bald man smiled and beckoned us over. Mr Rafi'i introduced him: 'This is Mr Mousavi, and that's the grave of his nephew who died on the last day of the War. The great Creator saw fit to draw him to his breast . . .'

'Thanks be to God,' interrupted Mr Mousavi dutifully.

'I was explaining to Mr Duplex here,' said Mr Rafi'i, 'the reason why these boys went.' He turned back to me. 'Boys like Mr Mousavi's nephew – Amin, wasn't it? – were in love with justice and God. Right, Mr Mousavi?'

'That's right,' said Mr Mousavi. 'The last time I saw him, he said – it was the end of the War, we thought he'd been spared . . . he said he was sorry that God hadn't judged him worthy of martyrdom . . .'

'Mr Duplex,' Mr Rafi'i said, 'you must know it's an honour to be martyred; not everyone gets called.'

Mr Mousavi went on: 'God heard him and took him on the last day of the War.' His expression went dead. He was awed by the severity of God's kindness. I looked at the photograph of Mr Mousavi's nephew. A normal kid, with 1970s bouffant hair and a beard and a spiky shirt collar. I looked along the line of photographs, at his neighbours. They had the same confidence; God wouldn't let them lose.

Mr Rafi'i and I went back to the grave of the boy whose body hadn't been found. 'At the end of the War,' he said, 'some of the old soldiers volunteered to go back to the battlefields and try and find the bodies of the missing lads. They even crossed the border, into Iraq. They used their knowledge of the sites, and their memories of the battles, to find the bodies. Then they dug them out and brought them back to their families.'

'So they found him?' I asked, gesturing at the grave.

He nodded. 'Five years after he died, they found him – his father told me. His trench had taken a direct hit. The strange thing is, his face was preserved, perfect. There was no smell, either. You know, the body decomposes and produces a smell. There was none of that . . .' He looked at me closely. 'Do you believe what I am saying?'

I wanted to believe him. Perhaps. It was fantastic, no? I nodded vaguely.

He started to cough, weakly, like a kitten. His face had got redder. A sort of yellow scum had accumulated at the corners of his mouth.

I said: 'Shall we sit down?'

We walked towards the trees. Mr Rafi'i laid his packsaddle on a grave. As we sat down, I said: 'Among Christians, it would be considered offensive to sit on a grave or walk over someone's grave.'

Mr Rafi'i was breathing a little easier. He grinned. 'The soul cannot be sat on.'

Something made him remember a young seminarian, Hamid, during the War. 'He was sixteen years old. He was a good boy: pure! They were all pure, back then.

'I remember – such a fine looking boy! Like the moon! He had an accident – I've forgotten what it was – and his front teeth were smashed in. I said he should go to the dentist and get his teeth repaired, and he said: "I'm not going to bother, because I've been summoned."

'A few days later, we went out to try and get an idea of the enemy's strength in our sector. We were twenty-two of us, in a column. As we set out, Hamid kissed me on both cheeks. He smelt of cologne, and he'd put on clean clothes.' If you're going to meet God, there's a protocol to be followed.

'The Iraqis were on the heights above us. When we came under fire, we hit the deck, and Hamid was next to me. I noticed my leg was hot and I thought, 'I've been hit', but something stopped me looking down. I was afraid. Then, a few seconds later, I felt that my groin and stomach were also hot and wet, and I looked down and I saw I hadn't been hit. It was Hamid's blood. I looked at his face. He smiled, and slept.' Mr Rafi'i looked up at the sky, to the bending tops of the cypresses and pines. 'You have to be clean, to be a martyr.'

Anticipating my next question, he said: 'God didn't want a grizzly old sinner like me.' The sprite laughed with us.

I asked: 'What did the men do, before they went off to battle? Did they pray? Were they silent? Did they chatter to try and settle their nerves?'

'I remember, before one operation, my battalion was given leave to go to the nearest town, to the public bathhouse. Normally before an operation you do your martyrdom ablutions and ask God to let you come to him. At any rate, everyone went along to the baths – we must have been about four hundred people. And I got a shock, I can tell you, because the lads in the bathhouse started mucking around, splashing each other with cold water. Afterwards, someone told me you could hear the shouts and laughter from down the street.'

'Did you joke around and splash, too?'

'No, it's not correct behaviour for a cleric to behave like that. I washed myself quickly and left.'

'How many of the boys who went to the bathhouse are still alive?'

'There can't be more than a few dozen alive now.'

He got out a bottle of water from his pack and took a swig, making sure his lips didn't touch it. Then he handed it to me. He said: 'So, you've come to Isfahan to learn about the War?'

I said, 'I hope to come to Isfahan several times.'

'You should meet Hossein Kharrazi.'

'Who's that?'

'He's over there.' He gestured behind him. 'I'd take you myself, but I need to have my injection. With your permission, I'll be on my way. Just go over there, and ask where Kharrazi is. Everyone knows.'

'Does he work here?'

Mr Rafi'i smiled gently.

After he'd gone, I went over to the area he had pointed to. I started walking up and down the lines of graves, reading the inscriptions and trying to find Kharrazi. After a while, I came across a group of people standing in front of a grave. It was identical to the other ones, but had more flowers. Flowers from a shop, some wild flowers and some plastic ones, too.

Shortly after the Revolution, Iran's mainly Sunni ethnic minorities started to demand administrative, religious and cultural autonomy – even, in some cases, independence. In the north-east, the Turkmens of Turkmensara; the Arabs of the south-western province of Khuzistan; the Baluchs on the border with Pakistan; Kurds living adjacent to Turkey and Iraq; all tried, using violent and diplomatic means, to persuade Khomeini to dismantle the centralized bureaucracy that the Shah had bequeathed him. Before the Revolution, Khomeini had promised freedom for all. That was before.

The worst violence was the Kurdish violence. The Kurds had welcomed the Revolution because they hoped to influence its course; the Shah's departure would not serve them unless the new regime recognized their historic separateness. The Kurds have more ethnic and cultural affinity to the Persians than they have to the Arabs and the Turks, their other hosts in an ancestral home the size of France, but they have been taught by history to distrust them all. The Kurdish imagination is littered with memories of betrayal and misery, much of it self-inflicted.

There's no reason to suppose that Khomeini rejected the regional stereotype of Kurds as murderous and untrustworthy. A fear of separatism, as much as his hostility to left-wingers and liberals, had persuaded him, shortly after the Revolution, to set up the Revolutionary Guard. When the Kurdish violence broke, Bazargan and some of his allies favoured conciliation; Khomeini and the clerical hardliners wanted to send in the Revolutionary Guard. Hossein Kharrazi, along with some fifty friends and acquaintances, got ready to fight.

Kharrazi was the third son of a junior civil servant in Isfahan, and his father couldn't pay for him to take up the university place he'd won before the Revolution. As a conscript in the Shah's army, he'd spent part of his military service in the tiny sultanate of Oman, across the Strait of Hormuz from the southern Iranian port of Bandar Abbas. Iran had kept a military presence there since the mid-1970s, when the Shah's troops had helped the Sultan suppress a guerilla uprising. Kharrazi went AWOL when, the day after the Shah's flight, Khomeini called for a mass desertion from the imperial army. Compared to many of his friends, whose military experience amounted to lobbing Molotov cocktails at the former regime's police stations, he was a seasoned soldier.

To Kharrazi and his friends, putting down the Kurdish rebels was a religious duty. He and most of the other lads had joined the Revolutionary Guard. (In time, the Guard would grow, allowing Khomeini to reduce his dependence on the regular army; he suspected it of remembering the Shah with fondness.) By a process of informal election, Kharrazi and another local boy, Rahim Safavi, became leaders of the group.

At twenty-two, Kharrazi was older than most of the others; being bright and fervent, he was able to articulate their ideals. America, the mortal enemy of the new Islamic Republic, was trying to turn Kurdistan into another Israel. Saddam Hussein, who feared the new Islamic Republic, was helping. As Shi'as and Iranians, it was their duty to fight. If they were killed – as long as they had not actively sought death, but rather the glory of Islam – they would be martyrs and go to heaven. (The Qoran and the sayings of the Prophet made that clear.) If they stayed alive, and won, they would recreate the Imam Ali's perfect caliphate.

To the boys in Isfahan, and across the country, that seemed like a terrific deal. The new warriors were mostly poor boys; similar boys in other countries, on both sides of the Iron Curtain, would have been drawn to extremist politics. Many were illiterate. The Shah's rule had disoriented them; the elite had been devoted to money, while much of the rest of society continued to profess its old attachment to spiritual rewards. These lads had a penury of both. Now, wealth was being measured in ways that favoured them. You were rich if you enjoyed the favour of God and the Imam, if you were going off to Kurdistan for a grand adventure – a love affair with the Revolution. You were worth a million if your mother shed tears of dread and pride on your shoulder: 'God speed your return!'

There was a jackpot up for grabs: martyrdom.

A minority of the boys – the more thoughtful ones – conceived of heaven abstractly. It was a state of grace, God's mingling with the soul. (By contrast, hell was regret, a longing for divine favour that throbbed into eternity.) Most of the lads, however, thought of heaven as a mild spring day, where the heavenly facilities could be smelt or touched.

Kharrazi, Safavi and the lads took a bus to Kermanshah, more than one hundred kilometres south of the war zone. They commandeered a helicopter and went to Sanandaj, the capital of the province of Kurdistan. As they touched down, the airport was being mortared. According to Muhammad-Reza Abu Shahab, who went on to become one of Kharrazi's aides, 'We had no knowledge of military operations, or military theory. God was helpful and gave us false confidence; we thought we'd beat them easily.'

The Kurds made insulting, overweening demands. (They were showing that they didn't understand the Revolution, and put their own petty nationalism above the rule of God.) They boycotted the referendum on the new constitution. Meanwhile, some misguided or treasonous elements inside the government continued to promote a peaceful settlement. They came to the region for peace talks, bartering with the Islamic Republic, when they should have been killing and dying for it.

Kharrazi and the others hated the army and mocked its daintiness. (The army was known to balk at orders to bombard Kurdish villages.) But the army was essential to the struggle to put down the Kurds; the Revolutionary Guard were too few, inexperienced and ill disciplined to win on their own.*

Kharrazi's men carried cumbersome old automatic rifles. These had been supplied – with great reluctance – by the regular army. If they wanted heavier weapons, they were told, they would have to steal them from the enemy. Gradually, as they killed more Kurds, they picked up their Kalashnikovs and Uzis.

There were cakes, presented to the Revolutionary Guard by local girls, which exploded once the girl was out of view. There was the beheading of a convalescing Revolutionary Guardsman in a hospital ward. There was the rape and disembowelling of boys whose fathers supported the government. There were government sentries found with their genitals stuffed in their mouths. There was the fear of fighting in tall valleys far from home, of depending on Kurdish guides who might be leading you into a trap. No one said the Kurds fought like gentlemen.

I don't know whether Kharrazi's group perpetrated what I, or you, might deem an atrocity. A bullet in the head for a Kurdish Sunni – punishment for refusing an invitation to join the Shi'a faith? A bastinado for a shepherd who failed to disclose the whereabouts of an enemy patrol? A shelling for a village whose inhabitants had given the rebels bread? Abu Shahab and

---

* For an illustration of the rancour that destroyed relations between the army and Revolutionary Guard, you only need to consult the memoir of Colonel Sayyad Shirazi, the hardliner whom Khomeini appointed to put down the Kurdistan rebellion. Subjected to a grilling – unfairly, he thought – by pro-army government figures and some senior regular officers, Shirazi claimed to have responded thus: 'I'm ashamed to note that this meeting, which has been convened in the interests of the security of the Islamic Republic, did not start with the preamble, "In the name of God, the compassionate and merciful", and that not one verse of the Qoran was recited. I regard the meeting as so polluted and unclean, I fear that my whole being has been corrupted and spoiled by it.' Shirazi announced that he would repair to the shrine city of Qom, in the hope that he could purify himself.

others deny that such things happened. But the government's campaign was famous for its savagery. Kharrazi and his lads were fighting the enemies of God, and nowhere in Islam does God prescribe for them a gentle rehabilitation.

That was perfectly understood by Khalkhali, scourge of Hoveida and destroyer of the Pahlavi tomb. For a while, he was in the rearguard of the advancing government forces, dispensing justice. His trials would last ten minutes or so. Verdicts depended on his whim. There was no appeal.

A magnificent society was being created. The unit was its microcosm. Men addressed each other as 'brother'. Kharrazi and his aides were obeyed because their authority, everyone assumed, came from God. When there was a shortage of food, Kharrazi would pretend he wasn't hungry and give his share to the younger lads. He took guard duty like everyone else, and went on dangerous reconnaissance missions. 'No one missed the cities,' one of his men told me, 'because they were still full of sin; here, in the fields, we were fighting alongside God.'

Kharrazi taught them:

No drop of liquid is more popular with God than the drop of blood
    that is shed for him.
The best deed of the faithful is fighting for God.
Participate in holy war, so you will be happy and need nothing.
One hour of holy war is better than sixty years of worship.
The wives of those who have gone to war must be respected and treated
    as inviolable.
An ideologically pure army is better than a victorious army.

They didn't realize it at the time, but it was all a preparation, a rehearsal for a grander struggle.

Even before the Revolution brought Iran's Shi'a clerics to power, Iraq's Baathist state had been suppressing its own discontented Shi'as. Although they constitute a majority of Iraqis, Shi'as had been underrepresented in the dictatorship that Saddam helped set up in 1968. They had been alienated by its espousal, variously, of secularism and Sunni nationalism. Saddam, who had been vice-president since 1968, but overshadowed the president in influence, regarded the new Iranian regime as a challenge – one that he might be able to turn to his advantage.

Before the Revolution, Iran and Iraq had been rivals, but the Shah had enjoyed the advantage. Being Arab nationalists, the Baathists had been furious when Iran, supported diplomatically by Britain and America, occupied three islands in the Persian Gulf in 1971. In response, Iraq expelled some seventy-five thousand Shi'as of Iranian origin and allied itself to the Soviet Union. But the Iraqis and Russians never built up the kind of intimate relationship that the Shah enjoyed with the US. In case of war, Iraq's elderly Soviet kit would be no match for Iran's expanding arsenal of sophisticated American weaponry. Access to US armaments was Iran's reward for being the 'third pillar' – the other two pillars being Israel and Saudi Arabia – of America's anti-Soviet policy in the Middle East.

In 1974, with American support, Iran intervened to help the Kurds of northern Iraq revive their intermittent insurgency against the government in Baghdad. As the likelihood of a destabilizing war between Iran and Iraq increased, so did international efforts to broker a lasting peace between the neighbours. In 1975, in Algiers, the Shah and Saddam signed an accord that required their respective governments to stop interfering in each other's domestic affairs, and which ostensibly settled outstanding border disputes. Iraq conceded to Iran part ownership of the strategic southern reaches of their fluvial border, the Arab River. The Shah, and the Americans, ended their support for the Kurdish revolt.

Four months before the Revolution, the Iraqis complied with the Shah's request that they expel Khomeini, who had been living in exile since 1965 in the Iraqi shrine city of Najaf. He went to Paris, which turned out to be a far better place from which to organize a revolution. On his return to Iran, he made it clear that he wanted political Islam to spread. In Iraq, thousands of Shi'a clerics, who remembered Khomeini from his period of exile, agreed. Ayatollah Muhammad Baqer al-Sadr, the most prominent of them, sent Khomeini a telegram in which he predicted that 'other tyrants' would also meet their reckoning. The Baathists put him under house arrest.

In June 1979, Saddam seized absolute power. He stepped up repression of militant Shi'a groups that Iran was arming and training. As the revolt in Iranian Kurdistan intensified, he took a leaf out of the Shah's book, providing the insurgents with cash and arms. (The Islamic regime started supporting the remnants of the Iraqi Kurdish group whose rebellion had been crushed in the wake of the 1975 Algiers Accord.) Saddam also armed secessionist groups in the Iranian province of Khuzistan, which sits on most of Iran's on-shore oil and gas.

In the first half of 1980, hostilities became overt. An Iran-backed Shi'a group made an unsuccessful attempt on the life of Tariq Aziz, Saddam's foreign minister. Iraq bombed an Iranian border town and expelled more Shi'as of Iranian descent. When he learned that al-Sadr was planning to visit Tehran, Saddam had him executed. The following month, Iran foiled a coup attempt sponsored by exiled monarchists. The death of the deposed Shah deprived the monarchists of their figurehead. If the Baathists wanted to bring about Khomeini's fall, it was clear that they would have to push him themselves.

In 1979, Egypt's Anwar Sadat had become the first Arab leader to make peace with Israel, earning the gratitude of the US and the opprobrium of millions of Muslims. With that deal, Egypt was deposed as unofficial leader of the Arab world. A show of force and determination, Saddam calculated, would make the position his. He wanted to suck Khuzistan into his sphere of influence, and to have a say in the composition of a post-Khomeini government in Iran, a government that would be friendly to Iraq. Iraq's state television was to describe the Iranians as 'flies'; the hatred, amply requited, of Arabs for Persians may have facilitated his decision to attack.

For the first time since the Baathists assumed power in 1968, it seemed as though Iraq could win a war against Iran. The revolutionary regime was split between moderates and hardliners. Although the ethnic rebellions, with the exception of the Kurdish one, had been put down, other armed groups, some of them leaning to the left, seemed to be preparing to come into conflict with the new clerical establishment. The Iraqis, Saddam was advised, would be welcomed as liberators by the Arabs of Khuzistan. In the event of determined aggression, Khomeini's republic would collapse.

Assessment of the two countries' respective military capabilities suggested that Iraq had closed the gap on its neighbour. Executions, desertions, purges and reassignments had cut Iran's army by 40 per cent. The US Embassy hostage crisis, along with Iran's hostility to the 1979 Soviet invasion of Afghanistan, had isolated the new regime from the world's biggest arms producers. On the eve of war, about 30 per cent of Iran's land force equipment, and more than half its aircraft, were not operational. Iraq's armed forces, on the other hand, kept on growing.

On 22 September 1980 Iraq invaded Iran. The land offensive, launched at four points along a seven hundred-kilometre border, was strikingly similar to an exercise that had been devised forty years before by British instructors at the Baghdad War Academy. An inefficient command structure, excessive caution and unfamiliarity with combined arms operations slowed the advance. After four days it came to a temporary halt.

For the first two days of the War, the advancing Iraqis were not met by any large unit. Iran's mobilization, when it finally got underway, was calamitously managed. It took one division six weeks to get from a base in eastern Iran to the theatre in the west. Many volunteers who went to the front were armed with Molotov cocktails. A plan, predating the Revolution, for the Americans to computerise Iran's spare parts inventories, had not been completed. The Iranians didn't know what they had in their stores.

❖   ❖   ❖

It's a few weeks into the Iraqi violation. Saddam's expectations have been confounded. Rather than divide them, the invasion has united normal Iranians; they're rushing to enlist in a kind of euphoria. The Iraqi advance has been slower and more costly than anyone expected. The Arabs of Khuzistan have reacted sullenly to their Iraqi 'liberators'. In a couple of weeks, when the front stabilizes, the Iraqis will have overrun more than ten thousand square miles of Iranian territory, including a third of Khuzistan, but only one important Iranian city, the port of Khorramshahr.

It's a cold day, and the Imam is sitting on a dais, underneath a sign that reads Allah. The men in front of him, most of them wearing military uniform, are crying. They're crying because their Imam is praising them and they consider themselves unworthy of his praise. 'I feel admiration,' he's saying, 'before these smiling celestial faces, before these heartfelt sobs.' The fighters, killers of Iraqis, convulse, tears pouring down their faces. 'I feel insignificant,' the Imam goes on. The weeping reaches a crescendo.

He's the greatest communicator. He understands television instinctively. 'In the Name of God, the Merciful, the Compassionate,' he starts his speeches, and then . . . silence. Fifteen seconds, or twenty: the Imam looking at you, through you, and the hairs on the back of your neck rise. His frail head slightly bowed, thick black brows like guillotines about to fall.

Marvel at his contempt, his contempt for Saddam's accounting of power and advantage, for the unmanliness of his assault. When the Imam talks to his people, it's without the histrionics of the actor Saddam, or Carter's wheedling. You learn to love, and fear, his inviolable monotone.

Like the imamate of Hossein, or Ali, his leadership is supranational. When he speaks, the world listens. (Before, when Kennedy and Nixon and Carter spoke, the Shah listened.) America, mighty America, quakes. As he addresses the people, the Imam inlays the War into the marble of Islamic endeavour. When he finishes, you realize it's impossible – morally, logically, physically – for Iran to give in.

'The difference between our army and theirs,' he says, 'is that ours is constrained. For our army, it's Islam that lays down responsibilities, whereas the other side has a free rein. They launch their shells and their ground-to-ground missiles . . . and they destroy an entire city. And they get congratulated. Our men don't do that. They can't. They won't.'

There will be no compromising the principles of Islam. How, then, can there be compromise with Saddam Hussein? What use is it to live, unless God is smiling and your conscience is at peace?

Having invaded, and got bogged down, Saddam is in the mood to settle. His first negotiating position: complete control of the Arab River, autonomy for Arabistan (that's Saddam's name for Khuzistan), and some tinkering with border areas. Impossible for the Iranians to accept, but a basis, some people think, for discussions that could go somewhere.

Now look at the Imam's (strictly rhetorical) counterproposals: Saddam's resignation; the surrender of all Iraqi arms to Iran; the handing over of Basra, Iraq's vital southern hub, to Iran . . . the Imam enjoys delivering these insults, these unconscionable conditions. They show the completeness of his contempt. Saddam can only be rattled by his placid fury.

'What motive,' the Imam asks, 'did he have in doing – without studying the subject, without understanding what the consequences would be, without taking into account our people – what a few devils like himself, whispering into his ear, told him to do? . . . What is his motive, rushing from pillar to post and inviting us to make peace with him?'

The Imam's questions aren't meant to be answered.

'How can we make peace? With whom? It's like someone telling the Prophet of Islam to go and make peace with Abu Jahal. In the final analysis, that's not someone you can make peace with.' Abu Jahal was Muhammad's uncle. He planned to have the final Prophet of God assassinated. He's the only one of God's enemies wretched enough to merit a verse in the Holy Qoran.

Khomeini draws himself up, pulls his heavy brown gown of camels' wool around him. It's a cold day. He's frail, elongated, monochrome in his white beard and black turban. He berates the arch-pipsqueak:

'You're the one who committed all these murders in your own country, and in ours, you're the one who had all those Muslims killed . . . Now! Imagine that our president and our parliament and our prime minister sit down and give you the time of day, and say: "Come in the name of God: the Arab River's yours, just leave us alone!"'

Khomeini, chuckling inwardly at Saddam's naivety: 'Is that what it's all about?' Across Iran, in villages and small towns, the people, looking at the TV, know that it's not.

At the end of our lives we must compile a log of our activities and present it to the authorities. Points are totted. Heaven, purgatory or hell; you go to one, and your performance on earth determines which. If we let God down in this world, he'll catch up with us in the next. Where's the gain in that?

'How are we to answer the downtrodden of the world, and what are we to say to the people of Iraq? If we get a missive from Karbala, and it says: "What are you doing, making peace with a person who killed our holy scholars, who jailed our intellectuals . . . ?" What peace does that leave us with?'

Here, the Imam is laying out the second big responsibility of the Muslim – to the community at large, to the oppressed. 'The question's one of religion. It's not one of volition. Our dispute is over Islam. You mean we're to sacrifice our Islam? What . . . Islam is land?'

No, Islam is not land.

'We shouldn't imagine that our criteria are material, or define victory and defeat in terms of what is organic and material. We have to define our objectives in sacred terms, and define victory and defeat on the holy battlefield . . . even if the whole world rises against us, and destroys us, we will still have prevailed.'

(This is just as well. The Gulf States and Jordan; some western European countries; several members of the eastern bloc; they're helping Iraq, militarily, diplomatically, morally. In Resolution 479, which calls for a cease-fire, the UN Security Council didn't even name Iraq as the aggressor!)

Iran is alone, like the fulfilment of a prophecy. The Imam rises and the men shout: 'Khomeini! You're my spirit! Khomeini! The smasher of idols!'

❖   ❖   ❖

The day I returned from Isfahan to Tehran, I went from the terminal to the office of Ali-Reza Alavi Tabar. In 1997, Alavi Tabar had opened a newspaper that argued that the Islamic Republic should be reformed. In 1999, the judges, who had different ideas, had closed it down. Shortly afterwards, he'd started a second newspaper, with more or less the same staff and typeface. That, too, had been closed. Later, he'd opened a third newspaper, with the same staff and typeface, and a name that was facetiously similar to that of the first newspaper. And so on. A few weeks before I visited him on my return from Isfahan, a judge had banned Alavi Tabar's sixth newspaper, after eight issues.

Alavi Tabar was plump. He would talk about the sport he was doing: mountain walking, running and swimming. He said he ate only yoghurt and salad leaves for lunch. But he was puffy round the chops and his eyes were watery. When I first knew him, he trimmed his beard, rather than shaved it, for revolutionary grooming contends that shaving is a Western effeminacy. Later on, perhaps reflecting his alienation from orthodox thinking, he'd shaved his cheeks and jaw, sparing only a severely shorn goatee.

He was one of the few Iranians I knew who would admit it when he didn't know the answer to a question. He showed no interest in the journalistic use that I would make of the notes of our conversations, or what I would write about him personally. I'd once seen Alavi Tabar while I was walking in the street, driving his Paykan in a grey synthetic shirt, looking like a bureaucrat in the middle carriage of the gravy train.

His office was on the third floor of a building in the Square of the Seventh of Tir. He had no computer or books on shelves. He didn't smoke, which would have produced a lasting smell. When he went home at the end of the day, he left the dregs of tea, which someone cleared up. The impression of transience that I got from his office accorded with the predictions I'd heard; it was said that Alavi Tabar would soon be arrested.

On the day I got back from Isfahan, we joked that he was the kiss of death for any newspaper. He was tired of opening new ones. He was teaching management and planning at a public university. He'd written about sociology and philosophy. He knew Arabic and theology. There would always be books to write.

Alavi Tabar believed that the Revolution had been diverted from its proper course. Islam's social and political manifestations could get more sophisticated, as society itself got more sophisticated. The Revolution was getting a bad press now, but that was the fault of the opponents of reform. The Revolution had done much for Iran.

'Before the Revolution, you had traditional towns where parents wouldn't allow their daughters to go to school; they didn't want them being taught by men, in places where there was no *hejab*.' He was referring to the Islamic head covering for women, which Khomeini had eventually succeeded in making compulsory. 'After the Revolution, they brought in the *hejab*, and woman teachers, and suddenly all these traditional families started sending their children – with great enthusiasm – to school.

'After the Revolution, I was working in the Ministry of Education, and we wanted to give poor kids the opportunity to do things that only rich children had done before. Like swimming. Before the Revolution, no one could afford to swim, except the rich. But we gave swimming passes to kids who did well in exams, as an incentive.'

I asked how much influence the old egalitarian impulse still had.

'A few weeks ago, I visited a cultural centre that the Tehran municipality set up. I asked the young girls how many of them knew how to swim, and almost all of them put up their hands. A poor neighbourhood, you know. And my wife watched them swim, and said they swam well. You know, those girls are learning karate. They all play a musical instrument.'

'And do you think these experiences have had a lasting effect? I mean, is there a difference between them and their parents' generation?'

Alavi Tabar smiled. 'It's curious you should ask that. I asked those same girls how they would react if their father said one day: "Now, look here; you should marry, and I know the right chap."'

'And?'

'Almost all of them said they wouldn't accept. One of them said she'd take her father to court if he tried to force her to marry against her will.'

Alavi Tabar had two daughters and a son. They teased him about his dreary clothes. If there was an argument about which TV channel to watch, they ganged up, so they were three votes, and defeated his one.

'Is that being modern?'

'Modernity means having the freedom and ability to criticize.'

'Not the victory of the individual?'

'No.'

The triumph of the individual meant the end of society. Alavi Tabar loved society. He derived his security from being part of it. It's a way of not feeling alone.

'Those who say the Revolution rejected modernity are quite wrong. Before the Revolution, modernity was an exotic foreign tree, whose fruit we were importing. Some of it tasted good and some not so good, but the essential fact remained; we knew nothing about the tree, just the fruit.'

'And now?'

'What you have now, after the Revolution, is a movement towards a kind of modernity that's not imposed by someone else, but something that comes up of itself. It's stronger, more authentic.' Saying his beliefs made his tired diluted eyes bright and opalescent. 'In the West, you've forgotten that freedom has a price; we know the value of freedom and we're willing to pay the price.'

A man brought in two glasses of tea and put them on the table. There was no cake or fruit – in line with Alavi Tabar's diet, I guessed. It was the season for tart little Iranian grapes, and nothing gets you chubby more deviously than those tart little grapes.

When Alavi Tabar picked up his tea, I noticed a small plaster on the tip of his right index finger. He'd been soaping himself in the shower that morning and his finger had slid over the cavity just above his collarbone – and got nicked. Touching the cavity a second time, he'd withdrawn from it a mysterious shard.

'It was a piece of shrapnel,' he said. 'It entered my body fifteen years ago. I gave it to my wife.' He was smiling quizzically.

I thought: a bit of the War inside him and now it's out. He's the right age. You meet people like that all the time. For all their mellow indifference, you know they've seen and suffered – and perhaps done – unspeakable things.

I said: 'I've just come back from Isfahan. I was there to learn about the War. I was trying to find people who fought with Hossein Kharrazi.'

Alavi Tabar was from Shiraz, in the south, but he remembered Kharrazi. 'It was the end of Operation Kheibar, in 1984. He was the last person to cross the pontoon they'd thrown across the marsh, towards our lines.'

'What do you remember?'

'I was ahead of him on the pontoon.' He paused. It was an effort. There was always a chance that he'd mix this operation up with another; there had been so many. Slowly, he said, 'Yes . . . I was carrying someone who was wounded. I remember my kidneys were hurting from the weight . . .'

'How did you know it was Kharrazi?'

'Everyone knew him. He was a fine commander, brave. He could direct five brigades at one time, and he was only a young man. But they were scared of him, his men; when he got angry, they were terrified.'

'What happened on the pontoon?' My neck was tingling. It was the first time I'd heard Kharrazi's name outside Isfahan.

'He shouted at us to run on and get away as fast as we could, and I was looking back, because he was the last one onto the pontoon. I was worried for him, and the Iraqis were firing at us with a machine gun, and the bullets cut through his arm.'

I said: 'That must have been when he lost his arm.'

I remembered something that Alavi Tabar had said in an earlier conversation, about admiring the big military heroes. He'd liked Napoleon – until he read an account, translated into Persian from the Russian, of the Moscow campaign. He admired Wellington, Rommel, Montgomery. He'd liked the way Montgomery wore his beret, at an angle.

I said: 'Would you talk to me about the War?'

He said: 'It's been such a long time. I haven't talked about it . . . there are people and friends . . . I haven't remembered for such a long time.' There was a pause. His face had crumpled. His eyes were red. He said: 'It was strange . . . very strange.'

He shook his head and wiped his eyes. He couldn't talk. He was shocked to discover that a memory could still wind him. We sat for a moment.

He cleared his throat and said, 'You know Susangerd?'

Susangerd was about thirty kilometres inside Iran. It was on the river. It commanded the road south-east, to Ahwaz. During their initial advance, the Iraqis had gone past Susangerd, and it had been badly bombarded. In November 1980 they tried to take it and hold it.

'It was the beginning of the War and we found out they needed men and we rushed up from Shiraz to help. I'd joined the Revolutionary Guard. The gendarmerie in Susangerd had been scattered to the four winds. There was just a trench at the entrance to the town, and the defenders were in the trench, trying to keep the Iraqis back with old-fashioned weapons. When the Revolutionary Guard and Basijis entered the town, they used the same weapons. We had old American M-1's, and some Molotovs.

'We'd heard news that the Iraqis had taken the surrounding villages and were raping women, and I myself saw the bodies of a few women and it was obvious that they'd been raped and killed.

'We entered the city in six groups of twenty-two and met up with other volunteers. There were Azeri Turks from the north, Yazdis and Tehranis as well. Everyone was happy to see each other. Then, the Iraqis sent their tanks towards Susangerd and the tanks rolled across the trench at the entrance to the town, and they rolled over our heads and went in. There was nothing we could do so we withdrew into the town.

'I was with a friend called Hassan. He took a bullet in the arm and I tied up his arm with my puttee, to try and stem the flow, and threw him over my shoulder, and I remember I was joking, saying, "You're so heavy, you should eat less" – that sort of thing. There was a mosque in Susangerd that must have been about the only undamaged building in the whole town, and I took him into the mosque and lay him down so he could die. He was losing an immense amount of blood, and I thought his tendon had been shot through. I thought there wasn't much we could do to save the situation; eventually they would get us all. I told him to stay in the mosque, and pray. Then, I went off to make my ablutions. I thought I might as well die having said my prayers.

'I went to wash my hands, to try and get Hassan's blood off. Then I saw an RPG launcher lying there. I didn't know if it was ours or theirs. I'd seen them being fired in a demonstration, but had never fired one myself. I filled my rucksack with grenades, and walked to the door of the mosque, and another friend, called Amir, was with me. I don't remember when exactly he turned up. Anyway, I remembered photos of people firing RPG launchers and someone was always holding their waists for the recoil. So I said to Amir, "When I get ready to fire, you hold me by the waist."

'We got to the top of the road and I saw a Russian T52. You could take out T52s with an RPG. It took some time to get the grenade into the thing, and then I fired. And it blew up. For a moment, we didn't know whether I'd done that, or someone else. And then we went off, around the town, to get more tanks. We discovered that there was no need for Amir to hold me by the waist because the recoil wasn't that bad.

'I felt that God was present and that we were distinguished people, fighting for him. I started to lose my fear of being shot, and I knew that even if I was killed I had a lovely future ahead of me. I was praying as I fought. I prayed that the people in the tanks had been Baathists, not conscripts who'd been forced to come and fight. We knew about the Baathists, and we felt they weren't Muslims.

'I was wearing a khaki uniform and a shirt over my trousers, and army boots that were tied with elastic bands because we had no laces. I had a tiny Qoran in my pocket that got hit later by a bullet. I had no hat. There was an enormous amount of dust. Later, I realized my beard had solidified. I couldn't pass a comb through it. Also, I wasn't used to the sound of mortars. There was the sound of rifle shots and occasional machine-gun fire, and sometimes groaning. Whenever our lads were hit, they'd shout, "God is Great! God is Great!" and we'd go and tell them to be strong and pray, because God was with us. I was sad and exhilarated, by turns. My feet were rubbing and sweating inside my boots.

'Over the next few hours, I managed to hit eight tanks squarely, and one in its tracks, and later we took possession of that tank and had the tracks fixed and were able to use it. The Iraqis also had infantry, but they entered the city after the tanks, and we got them too. There must have been six or seven hundred of us, trying to get the Iraqis out of Susangerd, but of course I don't know for sure. During the battle, the commander of my battalion was killed, and so I took over, and became commander of one hundred and ten people. Eventually, our auxiliary troops arrived, and they were better armed, and the city was liberated, but it had been pretty much destroyed.

'Our group took around twenty prisoners. They were mature men, older than us and pretty well built, and we were only twenty years old and weedy. There was one of our lads, so skinny we used to call him the wasp, and he took ten of these enormous Iraqis prisoner with a gun that only holds five shells. They could have got the better of him, but they didn't have fire in their bellies. It was the Iraqi Presidential Guard who fought well; the conscripts didn't fight well at all.

'After the battle, I didn't say anything about what I'd done, but everyone asked me how many tanks I'd got, and Amir told everyone. That night people came and congratulated me, and kissed me, and they wanted to hear my version of events, but it seemed wrong to boast. I said a prayer of thanks, but quite a few of my friends had been martyred, and I prayed for them.

'At the end of the night, I suddenly remembered my wife and I decided to write her a letter and tell her what had happened. I started, "My Darling

Afsaneh", and then I thought it might fall into someone's hands and be read, and that I'd better avoid expressions of affection. I wrote two pages and then they called me and I left my notebook and later on I came back and found that it had gone.

'Oh yes! I forgot to tell you about Hassan. I'd gone back to the mosque earlier, during the battle, and he wasn't there, and I assumed he'd died and they'd taken away his body, and I found some bread and dates and ate them because I was terribly hungry. I said some prayers for Hassan, and I wondered how I was going to tell his mother. But when I came out of the mosque, I saw Hassan. I shouted: "You're alive!"'

❖ ❖ ❖

A few days after the invasion, Hossein Kharrazi commandeered two buses and took his men to Ahwaz, the capital of Khuzistan. It was the Revolutionary Guard's regional headquarters and the regional headquarters of the Mobilization of the Oppressed, the Basij. Khomeini had set up the Basij well before the War had started. It was to be an irregular volunteer force, commanded by the Revolutionary Guard. Khomeini said he wanted twenty million Basijis. Since they had few guns, their weapon would be faith.

The Revolutionary Guard gave Kharrazi and his men four hours' training. They saw a truck-mounted multiple rocket launcher, a Katyousha. (They weren't allowed to take it; Katyoushas were few and far between.) They were told to march south. After a couple of days, they reached Darkhoein; the Iraqis had occupied the atomic complex. One of Kharrazi's scouts warned him that enemy armour was advancing. Some of the lads had never seen a tank. They were dismayed when their bullets bounced off.

One of Kharrazi's men disabled an enemy tank with a grenade. Supposing themselves to be under attack by a well-equipped force, the remaining Iraqi tanks withdrew. Kharrazi gave the order to dig in, but no one knew how to dig in; there hadn't been much call for trenches in Kurdistan. The hole they dug – 'a grave', in Abu Shahhab's words – violated norms that have governed trench building since World War I.

The front stabilized and Kharrazi started going to Isfahan. He consulted other commanders on strategy against the Mujahedin, which was continuing its civil war against the Islamic Republic. He visited Ayatollah Taheri, the Imam's representative in Isfahan, who was supervising the war effort there. He toured mosques that had been turned into recruitment stations. War mothers pickled vegetables and knitted underclothes – supplies were taken to the front in trucks lent by bazaar traders. Isfahanis still tell the story of an impoverished peasant woman who travelled all day from her village to a collection point in the city, where she donated an egg to the war effort.

Mr Zarif was fifteen, too young to join up but old enough to know what he was missing. His father, despite being enthusiastic about the Revolution, forbad him from going to the front to wangle his way into Kharrazi's force. (Many middle class parents felt this way.) Mr Zarif made plans of his own and his mother was a reluctant accomplice. One night, Mr Zarif slipped out of the house. Early the next morning, he phoned his mother from Ahwaz and asked her to persuade his father that he'd had no other choice.

In Ahwaz there was joy and chaos. Basijis roamed everywhere. Some were elderly seventy-year-old holy warriors, a bit confused, doing their bit. There was a Revolutionary Guard unit for women; the volunteers were being taught how to fight while wearing a chador. In the Shah's time, the Revolutionary Guard HQ had been a golf club patronized by Americans working at the Darkhoein atomic complex. The clubhouse had been renamed the Barracks of those awaiting Martyrdom. Men who might once have been caddies or flunkeys issued orders and prayed on the floor of the bar. The music system played laments for the Imam Hossein.

Each Revolutionary Guard battalion was assigned a cleric. They answered the men's questions on everything from martyrdom to daily observances and instilled great bravery in them. They encouraged the Basijis to wear the *kaffieh*, lest they forget that, even if Saddam was a considerable fiend, he was only a claw of a tentacular, Zionist beast. Mr Zarif heard rumours that some autonomous Iranian militias were counterattacking. One of them, it was said, was led by Sadegh Khalkhali.

As Mr Zarif had feared, Kharrazi's force turned him down. He identified a commander who had a reputation for flexibility and shadowed him, holding out his enlistment form for the man to sign. For a week, Mr Zarif barely left the commander's side. He gate-crashed important meetings. (No one asked what business he, a fifteen-year-old civilian, could have in such meetings.) Eventually, the commander relented and signed his form, but not before adding an infuriating proviso: ordnance.

The day they issued his dog tags he had another stroke of luck. Standing by the side of the road, he heard the roar of a motorcycle and a familiar voice: 'Boy! What are you doing there?' The voice and the bike belonged to a well-known PT teacher from Isfahan. He was setting up a sabotage unit. He was short of men and agreed to rescue Mr Zarif from ordnance. Mr Zarif was sent to a village near Darkhoein for thirty days' training. He found his speciality: television mines.

They had three legs supporting what looked like a screen and they were detonated with a device that the defenders held, sitting in their trenches. You laid them in front of your own lines, facing the enemy. Each mine dispatched about two thousand pieces of shrapnel.

Soon, however, the Iraqis learned to spot the slightly luminous television mines in the dark and they turned the screens round. After several Iranian defenders had received a mouthful of shrapnel, the saboteurs were ordered to stop laying them. But Mr Zarif came up with a solution. When laying the mine, he suggested, the saboteur should rub the screen with soil. That made the television mines all but invisible. After some successful trials, the television mine was brought out of retirement.

One of Mr Zarif's other jobs was to support commandos who carried out nocturnal missions behind enemy lines. They were only a year or so older than him. They carried knives and grenades and yelled 'God is Great!' while attacking, which terrified the Iraqis. Mr Zarif admired their swagger. One group delayed their return to their own lines until they had used the Iraqi showers and dined on the enemy's superior tinned food. Sometimes they came back in a hurry and Mr Zarif would have to give them covering fire.

Night was a time of fear, exacerbated by the exhaustion of being on guard duty in the forward trenches. Out in the darkness there might be barefoot Iraqi commandos, waiting for you to doze off so they could creep forward and garrotte you. The way to avoid this was to stand so your chin rested on the end of your gun barrel; if someone attacked you from behind, the barrel would get in the way. But this made Mr Zarif and the other lads nervous. Their safety catches were off. One false move – your jaw was in bits.

Soon, Kharrazi's force was up to company strength, about one thousand men. By early 1981, it was brigade-size: around three thousand. (A few months later, it would become a division, which is made up of three brigades.) Kharrazi named his force after the Imam Hossein.

Mr Zarif continued to be humiliated – and protected – by his age. If his name came up when the lads were drawing lots to decide who would participate in a dangerous mission, they'd invent some technical flaw and start again. Mr Zarif's commander once told him that a major operation had been cancelled and suggested that he return to Isfahan on leave. Mr Zarif got news of the operation when he reached home.

It was the time that some Basijis started to claim they'd had visions of the twelfth Imam. It was a tantalizing idea; he's the last of the Prophet's direct male descendants and the last infallible human being on earth. Although he was born in the ninth century, the Owner of Time never died. He's among us, in disguise. When he eventually reveals himself, this will presage an era of justice and truth.

Wise comrades warned Mr Zarif to resist the fashion for 'visions' – many commanders realized that they were bad for morale. You got lads sticking to a comrade under fire. The comrade had seen the twelfth Imam, and thus had an 'aura'. If a 'seer' were martyred, other men would rip pieces of clothing off his corpse and keep them as relics. Hossein Kharrazi had no patience with this indiscipline. He would beat the 'seer' until he recanted.

From talking to the men who fought under him, my impression is that Kharrazi cared more about the average Basiji than many commanders; a fruitless death, he maintained, didn't count as martyrdom. His men, although conspicuously brave, didn't fight with the recklessness that was ascribed to volunteers from Tehran. Kharrazi encouraged the growing number of Basijis under his command to go back home to Isfahan during lulls. They returned to the front when an offensive was in the offing. But that didn't affect their effectiveness as a fighting force. It may, in fact, have enhanced it. When Iran eventually started to recover its territory, the Brigade of the Imam Hossein was part of every major operation.

❖ ❖ ❖

Abd-Hassan Bani-Sadr had been elected to be president in February 1980, three months after the fall of Bazargan's provisional government. He'd got 75 per cent of the votes cast, but that wasn't a fair gauge of his standing. Events had conspired in his favour. The Imam had let it be known he didn't want a mullah standing for the presidency; to start with, he was concerned that the clerics not have a monopoly on power. The only genuine revolutionary among the front-runners was disbarred because his father turned out to have been an Afghan national.

Bani-Sadr was known to speak French, a villainous tongue. He'd studied in Paris – which everyone knew was the capital of depravity – with a Marxist sociologist! He hadn't been in Iran during the Revolution; he returned with the Imam. He had a shifty, left-winger's moustache and wore suits with fashionable lapels. He showed more interest in the economy than was proper. He thought a good deal of his own abilities. He enjoyed the company of Western peacemakers: Olaf Palme, for example (another lefty – a libertarian Scandinavian). He seemed to think that the Revolution could be cooled, or slowed down; he entertained the possibility of striking a peace deal with Saddam. During an address, he asked people to make the coming year 'the year for a restoration of order and security'.

No! That would be no fun at all!

As president, Bani-Sadr had theoretical command over both the regular and volunteer forces, but the Revolutionary Guard didn't listen to him. His generals allocated to the Basijis rusting rifles that dated from the reign of the ex-Shah's father. According to one of Kharrazi's men, it wasn't unknown for regulars to let the air out of their own tyres – 'to give them an excuse for not advancing'. Rumours of treachery circulated. During one engagement, it was said, the regular army had left trapped Revolutionary Guardsmen to die. There were reports that Revolutionary Guardsmen deployed behind the army had shot regular soldiers who were trying to flee.

Bani-Sadr erred towards rationalism. A Child of the Enlightenment, he propagated the fallacy that expertise and competence were more important than an attachment to revolutionary ideals. He didn't understand the Basiji slogan 'Blood will defeat the sword'. (This meant that, if a pure-hearted Iranian died on the road of God, his blood – his sacrifice – would defeat the great array of countries that stood against him.)

The Basijis repudiated Bani-Sadr's argument that the War could be won only if military sales by Western powers to Iran were resumed. That would be to imply that Iran remained in thrall to outsiders – that the Revolution was nothing more than a change of clothing. It was a mendacious suggestion, consistent with Bani-Sadr's advocacy of a quick solution to the US Embassy hostage crisis. Both positions were designed, the volunteers felt, to advance Bani-Sadr at the expense of all else.

For all his ponderous intellectualism, Bani-Sadr had a pragmatist's eye on the twelve billion dollars of Iranian assets that had been frozen in American banks, and the half a billion dollars' worth of military spare parts, paid for by the Shah, whose delivery had been held up. The terms of a putative deal to free the hostages would naturally include provision for the unfreezing of these assets. Bani-Sadr hoped to persuade the US to lift its embargo – America's allies would follow suit. He may have had ambitions to institute a new bilateral relationship, on a basis of honourable equality.

To Bani-Sadr's opponents, anything other than enmity for America was unthinkable. The Imam had called the US the 'Great Satan'. Hardline clerics were determined to prevent the president from striking a deal with the US that would gain prestige for him and a supply of weapons for the army. They suspected him of building up the regular forces so they could be used against them. Far from the enlightened revolutionary he purported to be, they whispered that Bani-Sadr was an ally, perhaps an agent, of the US.

In April 1980, Carter had ordered US commandos and local CIA agents to try and rescue the hostages. The operation was aborted after a helicopter crashed in the desert, six hundred kilometres south-east of Tehran, with the loss of eight American lives. In their haste to get away, the Americans left behind one million dollars in Iranian currency. It was an excuse for Sadegh Khalkhali to re-emerge into the limelight; he made a brief and grotesque contribution to America's misery, appearing on the networks fingering the bodies of charred US servicemen.

On 4 November – every Basiji knew the date – Carter would be up for re-election. If the hostages were free, he would get the credit and might well be returned. If, on the other hand, they were rotting in Iran, the electorate would punish him. Negotiations, using intermediaries, were going on between the two governments, but a deal was distant.

Iran's hardliners enjoyed the idea of deciding the Americans' election results. They wanted to pay them back for 1953, when the CIA had engineered the coup that allowed the Shah to entrench his despotism. They had no intention of helping Carter, whom they loathed for allowing the Shah to take refuge in America after his flight from Iran. They would rub Carter's nose in the dirt, and make anti-Americanism the core of revolutionary foreign policy. They would prevent Bani-Sadr's army from getting stronger.

Did Khomeini intend all along to abandon Bani-Sadr, his old protégé? Was the Imam 'turned' by his advisers? Even now, few people know for sure the extent to which Khomeini was all-powerful or manipulated. But it's not un-Iranian to allow people to think that nasty events happen a little over one's own horizon, to hint rather than stipulate, and to avoid precision for fear of attribution. The Imam frustrated Bani-Sadr. He relieved the president of responsibility for setting Iran's conditions for a resolution of the crisis. He gave this responsibility to parliament, which was dominated by hardliners who were in no hurry to set terms. Bani-Sadr's service chiefs were warning of military collapse unless the supply of arms resumed. The president's opponents enjoyed his frustration.

There was no agreement before the election; Carter was defeated. Then, a few days before Ronald Reagan's inauguration, a strange deal was struck. Although Iran's parliamentarians professed to loathe America, they set conditions for the hostages' liberation that favoured the US. They agreed that Iran would pay its outstanding loans to US banks in full, rather than just the interest that had accrued on them. They consented to entertain hundreds of legal suits by American companies that claimed to have been disadvantaged by the crisis. They left dormant the question of the frozen military assets. On the day of the inauguration, the hostages were released. Rumours of foul play have persisted to this day.*

---

* In his memoir, Bani-Sadr says, 'I have proof of contacts between Khomeini and the supporters of Ronald Reagan as early as the spring of 1980.' (That was months before the American election.) These 'contacts', Bani-Sadr says, prepared the ground for a deal

The conclusion of the hostage crisis was the beginning of the end for Bani-Sadr. A few days later, he ordered several counteroffensives against

---

whereby the Republicans got assurances that the hostages would not be freed until after polling day, making Reagan's election a formality. In return, says Bani-Sadr, the Republicans promised to help Iran circumvent the arms embargo that Carter had put in place.

During his 1988 presidential election campaign, Vice-President George Bush was challenged with claims that he and William Casey – who had been Reagan's campaign manager and went on to be his CIA chief – had struck a deal with Iranian hardliners to subvert the 1980 election. In the end, the source of these claims, an arms dealer called Richard Brenneke, turned out to be unreliable. The issue was dropped and Bush won the election.

Three years later, Gary Sick, who had been Carter's principal aide on Iran, revived the theory of a deal. His subsequent book, *October Surprise*, gave details of meetings, in 1980, between Iranian hardliners and Republican Party officials, Casey included. There were four meetings, says Sick – two in Madrid and a further two, to finalize the deal, in Paris. In return for helping Reagan come to power, Sick writes that Casey prevailed upon pro-Republican elements inside the CIA to encourage Israel secretly to supply spare parts to Iran.

Sick concedes that his story is very difficult to prove. Some of his sources were arms dealers and intelligence agents, speaking anonymously. The calibre of those who consented to speak on the record is questionable; before the story broke, several of them had been indicted or were the subject of US federal investigations. As Sick told me at the beginning of 2003, 'I didn't find a smoking gun.'

But all attempts to clear the alleged principals have failed. In 1990, a US court tried but didn't manage to find Brenneke guilty of perjury when he maintained, in an unrelated court case, that Bush, Casey and others had attended the Paris meetings. (Sick himself doubts whether Bush attended the meetings.) Bush denied any involvement in illicit dealings with Iran, but carefully stressed that 'all I can speak for is my own participation or lack thereof'.

In 1993, a US Congress task force rejected Sick's allegations. But Sick sticks to his guns. The investigating committee, he says, 'didn't follow leads and they didn't go down avenues that, they feared, might lead them to places they didn't want to go'. In the end, says Sick, the committee 'pretty much invented an alibi for Casey'.

If there was a conspiracy, the only way we will know is if the Iranians involved give their version. (Most of the alleged Iranian principals continue to occupy high office; they would have nothing to gain from talking.) According to Sick, Yasser Arafat told Carter that the Republican Party had approached him, at the time of the crisis, with a view to opening contacts with the Iranians about the hostages. The Palestinian leader, Sick says, told Carter that he would provide copies of notes taken at the time, but didn't. It is odd that none of the alleged American principals has so much as threatened Sick with legal action.

In 2003, Ayatollah Hossein-Ali Montazeri, who was close to Iran's centre of power at the time, told me that he had no 'documentary evidence' to support Sick's theory, but did not refute it. Montazeri added that it had been the opinion of many, 'including myself, that it would be in the interests of the Islamic Republic for the hostages to be released before the 1980 elections. I put this view to the late Imam, but he didn't accept it.'

Iraq. Being premature and ill-conceived, they failed. (In one, the biggest tank battle since the Six Day War, Iran lost well over one hundred tanks.) From then on, the army was rarely permitted to launch offensives without subordinating itself to the Revolutionary Guard and Basij, and the volunteer forces grew in stature. This was reflected in the spring budget; it was generous to the Revolutionary Guard and Basij, and less so to the army.

In the past, Khomeini had considered Bani-Sadr to be the representative of a legitimate thread of revolutionary thought. Now, he was shown evidence, drawn from material found and collated in the US Embassy, purporting to show that Bani-Sadr had once been considered suitable for recruitment by the CIA. Parliament began stripping the president of his powers. As his position became more precarious, and his supporters were arrested, Bani-Sadr foolishly allied himself with the Mujahedin, which was on the verge of launching a civil war against the regime. Eventually, he invited the people to 'rise' against the 'tyranny' being unleashed against him and went underground.

Cue for Sadegh Khalkhali to move briefly upstage; who can forget his Puckish appearance on the parliament balcony as the deputies impeached the president, putting his chubby fingers around his own neck while the crowd roared for Bani-Sadr to be hanged?

Bani-Sadr's flight abroad gave rise to an explosion of violence by opposition groups, of which the Seventh of Tir attack was the most lethal. In August 1981, Rajai, Bani-Sadr's successor as president, was blown up and killed. These attacks provoked a spasm of retribution. According to government figures, Iran executed more than two thousand members and sympathizers of armed opposition groups between June and September.

The hardliners felt vindicated by Bani-Sadr's treason, but they may also have been temperamentally incapable of working with someone who had taught at the Sorbonne. In the words of one senior revolutionary official: 'While Bani-Sadr lived in a gilded cage in Paris, we, the disinherited of the earth, the real Revolutionaries, rotted inside the Shah's jails. While Bani-Sadr gave interviews to *Le Monde* and *Der Spiegel*, our jailers made us drink our own urine and violated our sisters before our own eyes.'

It is hard, despite the unfair way he was treated, to feel much sympathy for Bani-Sadr. He may not have been as interested in democracy as he said he was. He gave every impression of supporting the 'cultural Revolution' that led to the expulsion of leftists from university campuses and to the purge of thousands of teachers and students. His books are self-serving screeds. He quotes his enemies favourably when they compliment him, and unfavourably when they don't. He disparages mullahs and expresses surprise at Khomeini's authoritarian tendencies. But this is the same Bani-Sadr who vaulted to power on Khomeini's back – and could not resist boasting that he enjoyed a 'father–son relationship' with the Imam. How come he failed to see it coming?

Bani-Sadr's enemies made much of reports that he had fled Iran wearing a chador. In his memoir, he primly refutes this, mentioning in passing that he was in military uniform. But Bani-Sadr's famous moustaches were gone when he reached Paris, where he has continued to live, his revived whiskers quivering resentfully, to this day.

I asked Alavi Tabar about his dreams.

'One night at the front, when my wife was in the latter stages of pregnancy, I dreamed I was walking in a forest and I saw a lovely flower. I bent down to the flower and said, "*Salaam.*" The flower replied, "*Salaam*", and it had the voice of a little girl. I asked the flower, "What's your name?" And she replied, "Sara." The following day, I told my friends that my wife had given birth to a girl. When they heard about the dream, and my interpretation, they scoffed, but I went to the nearest phone to try and find out. My wife told me over the phone that she'd given birth to a girl. I said the girl's name should be Sara.

'Another time, I dreamed I was running somewhere and I leaned down to pick up my rucksack, which had fallen, and it exploded. Then I woke up. It so happened that a short time later, when we were entering an Iraqi trench, I saw a rucksack and leaned down to pick it up. Then I remembered my dream and, rather than pick it up, I examined it more closely. I saw that there was an explosive fuse attached to the rucksack. It had been booby-trapped.

'In another dream I remember from the War, I'm looking for someone in the dark and I can't get to him. In frustration, I hit the side of the wall with my fist, and suddenly a door opens and I see there's someone inside and he's injured. A little while after I had that dream, we attacked the Iraqi sector at night. As we were mopping up, I heard the sound of groans. But it was dark and I was in a trench, so I couldn't find the origin of the groans. I was tired and leaned up against what I thought was a wall, but it was made of cloth and I fell through. Inside, I saw there was a Basiji whose chin had been shot off. He couldn't cry out; he could only groan. I got him out and gave him a fireman's lift as far as the nearest first aid post. He was bleeding profusely and his blood got everywhere, even in my mouth.

'Sometimes, I really did feel that God was protecting me. One day, I was sitting down and I saw a rat and gave him an almond. Then, I went after him, just playing around, and a mortar shell landed exactly where I'd been sitting. The rat scuttled off and I was saved, but the last thing I wanted was for such stories to give rise to superstition. One day, a young Basiji asked me, "If you get martyred, do you mind if I keep your helmet?" I tore a strip off him for that.

'I had a friend at the front who was very mystical. His name was Habib, and he'd say that everything exists to praise God, and he was in love with God in an intoxicated way, and was always cheerful. Some days, the Iraqis would rain down mortar shells, but he was never scared. He saw something beautiful in everything. He even loved the Iraqis. He always used to say that he'd end up being killed in a canal and that they'd never find his body.

'He'd make fun of me, saying I was too rational – that I could never be a lover. He said you're Aristotle and I'm Plato. He loved mystical poems and he'd write them on my vests when they were hanging out to dry, so I'd be wearing a poem during the battle. He believed that he was there to perform a service. When he had nothing to do, he'd sweep the trench clean, or we'd wake up in the morning and find that he'd cleaned the plates and polished everyone's boots and oiled the guns and sprinkled water over the trench floor. One day, he walked twenty kilometres just to bring us water. He used to say that God isn't in the cities; he's here with us, at the front.

'One night he raised a stink when one of the Basijis killed a snake. He shouted: "What harm did that snake do to you?" Then, he spent half an hour machine-gunning Iraqis. I said, "Here you are, who raised such a fuss over a snake; now you're killing men." He said: "That snake did no one any harm, but these Iraqis have invaded my country and they've raped our women and, if I don't kill them, they'll kill more of us." He'd say that he killed Iraqis in order to deprive them of more opportunity to sin.

'I was always badgering him to get married. One day he seemed to relent, and said, "I'll get married after the next operation." I was pleased and said, "All right; after the next op we'll find a wife for you." But he said, "That's not what I meant. I meant that I'm going. God's got a fabulous wife lined up for me." And he was killed during that operation, exactly the way he'd predicted. Three times I went and searched for his body around the canal where he died, but I couldn't find it. It was exactly as he said it would be.'

❁   ❁   ❁

At the end of 1981, Iran started recapturing territory. Sayyad Shirazi had been put in charge of the regular army, and cooperation with the Revolutionary Guard and Basij increased. By the end of the spring, Khorramshahr was the only important bit of Iran still in Iraqi hands.

On 1 May 1982 the Division of the Imam Hossein was part of a formation that crossed the River Karun, the effective dividing line between the Iranian and Iraqi zones in southern Khuzistan. Landing on the western bank, the Iranians quickly advanced to the Ahwaz–Khorramshahr road, an important line of communication between Iraq's field units and the occupied port, and cut it off. A week later, the Iranians were ten kilometres from Khorramshahr.

The Iraqis had fortified Khorramshahr with earth barriers in the shape of arcs, and made fire zones by levelling buildings and vegetation. The Iranians sustained heavy casualties in the palm groves north of the city, where there were mines, barbed wire entanglements and booby traps, but the Iraqis' caution deprived their armour of the ability to manoeuvre. On 21 May, Shirazi committed the Division of the Imam Hossein against defenders on the corridor linking Khorramshahr and the land border. For a while, the advance was held up by a burning tank that the Iraqis had purposefully left on the road. Under heavy fire, Kharrazi appeared on a motorbike and directed efforts to remove the burning carcass.

On the morning of 22 May, some seventy thousand Iranian troops made their assault on the city. (They were faced by about half that number.) It took the Iranians about a day to break through the huge earth and sand rampart, almost three metres high, that extended in a semi-circle around Khorramshahr. Kharrazi's division was among the first to enter the city, and Iraqi resistance started to collapse. Almost immediately, the Iraqis began to move forces back across the Arab River. Iran was free and there was joy everywhere.

CHAPTER FOUR

Qom

The sun beats in a sky without cloud. Your mouth waters, you think of rain. At the entrance to the precinct around the shrine, pilgrims from other places in Iran and other countries that have many Shi'as – Bahrain and Pakistan, India and Lebanon – stand mumbling. They're greeting Fatemeh, sister of the eighth Imam, Reza. She's called the Innocent One, and she lies in the shrine. They gaze at the dome, a carapace over her purity. They kiss the door jambs, one after the other, and colds caught in Karachi spread to Beirut.

According to Mr Zarif, the dome over the shrine in Qom is considered especially fine; in Islam, there's an optimum in architecture, just as there's an optimum in behaviour. The dome is tulip-shaped and depleted. The authorities recently stripped its gold plate, which had been worn by wind and rain, revealing the copper underneath; an Isfahani firm has won a contract to lay fresh gold plate.

In the road outside, close to the gutter, there's the gravestone of a mullah. He reserved the plot before the thoroughfare was expanded. After he died and was buried, the municipality decided that he was a nuisance and moved him to another grave, far from the shrine, where he wouldn't be an impediment to traffic and commerce. A few days later, he visited an important ayatollah in a dream and informed him that he'd moved back to his original resting place, at the side of the road. They put back his gravestone and left him there.

Inside the entrance to the precinct, a wreck of a man is reciting devotional prayers in Arabic for an unlettered family. These prayers are the supreme homily for the Innocent One. It's impossible, the authorities attest, for a human being, using whatever language is at his or her disposal, to better it. The man claims to be a cleric, but he can't afford a turban and looks as though he sleeps rough. He encourages tips.

The shrine is preceded by an oval court that historical photographs show to have been enclosed by a mud wall; now, after extensions that include a clock tower, it's gaudy and pretentious. Inside the court is the cordoned-off grave of some mullah, which pilgrims have covered with birdseed. Pigeons, as fat and sleek as cats, are feasting; their only anxiety is the urchins who recently caught one of them and ate it roasted.

As you walk across the court, towards the entrance to the shrine, you step on a stone slab that announces this to be the resting place of some man and his wife. You feel envy that they, by dint of their wealth or position, were able to get a plot so close to the Innocent One.

There's a widow ahead, talking to a young mullah; a compact – a temporary marriage, sanctioned by Shi'a law – is being proposed. 'Excuse me, sir,' she responds to his furtive come-hither, 'would you be interested in doing a good deed?' (After a few weeks of frenetic good deeds, the mullah will return to his studies – with luck, no longer distracted by lust. Into the widow's purse, he will drop a month's rent on her rooms. All this without violating holy law.)

The ladies' entrance to the shrine is beneath an ostentatious portico on spindly columns, which supports two minarets. Much of what is glittery and oddly proportioned here was built at the beginning of the nineteenth century, by Fath-Ali Shah of that mediocre and feckless dynasty, the Qajars. Fath-Ali revered the Innocent One; when visiting Qom, it is said, he would dismount and take off his shoes outside the city limits and complete his journey barefoot. The ladies are standing underneath the portico, taking off their shoes, touching the columns. Bags protrude from under their chadors. The stalactites in the huge niche glare with mirrors.

You surrender your shoes in return for a plastic disc with a number and enter a big room made up of a rotunda encircled by a flat-roofed arcade. The dome is supported by art deco columns that taper inversely; their stylish bases, brackets shaped to look like lions' paws, and the veined marble floor, would be appropriate to the headquarters of a big corporation. Behind the curtain rail that divides the room, women are praying.

On this side, most of the men are facing Mecca. One of them, an old mullah, has imploring red eyes. He is calling on the Innocent One to pilot him through an emotional or moral peril – the death of his wife, perhaps, or a sin. You walk on, into a domed chamber, quite small, with dark walls flecked with silver arabesques. Under the marble lies Fath-Ali Shah. Vengeful revolutionary authorities have removed all outward reminder of him.

Now you're at the door to the sanctuary. It's a tight room with an overwhelming gold-plated sarcophagus and a pungent smell from the feet. You join the slow anticlockwise churn, fingering the barred panels and Qoranic inscriptions, while the sweat begins to pour down the sides of your face. A woman is praying for the Innocent One's intercession in the matter of her daughter's barrenness. The ceiling fans push the heat and smell into your face, up your nostrils. Women and men barge and reach past you, touching the beaten gold. Attendants urge you not to dawdle; they hold pink feathers for rhetorical effect and wear blue station masters' uniforms.

You spin off from the lugubrious revel into a modern prayer hall. Here lie the aristocrats of Qom, the great ayatollahs who turned a dusty shrine town into the seat of Shi'a learning. There's Alameh Tabatabai, the father of modern Shi'a philosophy, Araki and Golpayegani of the pre-revolutionary generation, and, beyond that, in a small room to himself, Borujerdi, Shi'ism's top cleric when Khomeini was coming into prominence. There's a newer stone near the entrance. That commemorates Ayatollah Shirazi, who died on bad terms with the regime, in 2002. To slight him, they put him in the corridor.

Having collected your shoes, you emerge from a second entrance. The sun is beating you in the face. A little girl is running, chasing pigeons up into the air. A troop of pilgrims migrates noisily across the burning flags. The voices and heat merge. There are dozens of mullahs in the courtyard, most of them moving towards the Grand Mosque. They take off their shoes at the entrance and put them in a plastic bag.

Inside the mosque, the tumult is muted. At least the air in here is moving, propelled by ceiling fans. Soon, you feel a little better, and look around this Islamic School of Athens. Qualified mullahs, most of them teachers, sit cross-legged, leaning against the columns, talking and rereading the text for today. The enormous room is filling, acquaintances are acknowledged with a hand to the breast and a slight bow. There are moon-faced Afghans with Chinaman beards, Iraqis from Najaf, a contingent of Azeri Turks from the northwest. One mullah nearby has a powder-blue frock coat, from which he withdraws a fob watch. A prim *seyyed*, or descendant of the Prophet – identifiable by his black turban – stares into the folds of his sleeve, then up at the ramshackle latecomer in his unironed gown, his face berry-red from sleep and opium.

Then you notice a bowed man who is proceeding like a barge through the assembly. He looks immensely old. The mullahs near the front clear a path for him. He climbs the lectern's two or three steps, and sits down, cross-legged, on a cushion at the top. He takes a sip of water from the glass in front of him, and says: 'In the Name of God, the Merciful and Compassionate.'

The lesson begins.

One day in 1928, Reza Shah's wife or mother – versions differ – visited the shrine of the Innocent One. Inside, she inadvertently showed her face and was castigated by a senior ayatollah who witnessed the scene. The following day, Reza himself came to the shrine, accompanied by four armoured cars and about a hundred men. He showed his anger and contempt by entering the shrine without taking off his boots. Having identified the errant mullah, he assaulted him and, according to some accounts, gave him a horsewhipping.

Reza Khan, as he was known until he became Shah, came from a modest north Iranian family. He started his career as a Cossack trooper under the Qajars. At the turn of the twentieth century, Russia and Britain, while tirelessly reiterating their regard for Iran's independence, were effectively dividing the country in two; their interests dictated the form and character of the Tehran government. In the north, some security was maintained by a brigade – later a division – of Cossacks, which was commanded by Russians but was theoretically answerable to the Shah.

The Cossacks took varying approaches to the struggle between the proponents of a constitutional monarchy and the supporters of despotism. (Iranians won their first parliament in 1906, but it was dissolved, bombarded and otherwise disrupted before Reza came to power.) Reza fought both with and against the constitutionalists, and alongside the Imperial Russian army during World War I, when it countered Iran's pro-German proclivities. Thanks to his courage and initiative, he was promoted fast. He was given a moniker, Reza Maxim, because of his flair with the British machine gun of that name. There is little evidence that he progressed culturally, even after he was promoted to colonel. The Cossacks, Russian officers included, were notorious boors.

By 1921, lawlessness, rebellion and foreign encroachment had brought Iran to its knees, and Reza took his chance. With the connivance of British officers and diplomats – who feared that Russia's new Bolshevik leaders would profit from the chaos and invade – he took control of the Cossacks, marched on Tehran and toppled the government. The following day, Ahmad Shah Qajar was forced to appoint a new prime minister. Reza was rewarded with formal command of the Cossacks.

The Cossacks became the nucleus of Iran's first unified army and the base from which Reza came to power. He used the army to crush regional insurgencies and brigandage, and to bring order to the provinces. His popularity grew and he swiftly rose to the premiership. At the end of 1925, parliament abolished the Qajar dynasty – Ahmad had already scuttled off to Europe – and Reza Maxim became Shah.

I have a photograph of his investiture. It shows a big man – he was six feet three according to Vita Sackville-West, who saw him in person – in uniform, standing on the dais of the Tehran parliament, surrounded by mullahs. The mullahs are relieved; they are exercised by the fear that Iran's monarchy will one day be abolished, and replaced by the secular republicanism that Mustafa Kemal (later known as Ataturk) is pursuing in neighbouring Turkey. In this photograph, however, the new Shah is swearing on the Qoran to uphold the Shi'a faith. He has made a pilgrimage to

Shi'ism's holiest sites, in neighbouring Iraq. The mullahs expect Reza to protect Shi'ism and cherish its clerical representatives.

There is a second photograph, taken in the open air, the day after Reza's investiture. This time there are no mullahs. Winter sunlight falls on attendants wearing the traditional caftan and fez; their hands are joined before them in a gesture of servitude. Reza is scowling, seated on a cele-brated throne of ivory – part of his Qajar inheritance. He has a jowl like an anvil. He leans forward on the throne, impatient for the ceremony to be over.

At the beginning of his reign, many mullahs were taken in by Reza's ostentatious piety. During his spell as prime minister, he had welcomed clerics who had been expelled from the shrine cities of British-run Iraq and had settled in Qom. Two of the most senior of these clerics had declared obedience to Reza to be a religious duty. Once he was on the throne, however, it became clear that he regarded mullahs as an impediment to the modernity he wanted to impose on his country. Under the guise of Western-style reforms, he started to undermine them.

They were still the most powerful institution in the country. In the towns, they acted as judges and notaries; their seminaries received unquan-tifiable sums from believers. In rural areas, by virtue of their education, they might be asked by an illiterate farmer to conduct a correspondence with a government department. Pious men commissioned sermons from them to mark an auspicious day. From city bazaar to local market, the mullah's assent was needed for commercial contracts – otherwise, it was feared, these contracts might contravene Islamic law.

Such services provided the mullahs with an income and placed them at the centre of the community. But Reza envisaged a time when Westernized, qualified men (and perhaps women) would perform most of their func-tions, and Western-style laws, not Islamic ones, would determine the conduct of society.

There is no evidence that Reza was intellectually equipped to share Ataturk's agnosticism; he probably considered himself a respectable Muslim. But his assault on the clergy was as ferocious as Ataturk's. In the words of one historian: 'The sixteen years of rule by [Reza] can fairly be described as a period of intense hostility to Islamic culture and institutions;

what Western authors have approvingly called "reform" and "modernis-ation" was experienced by many – if not most – Iranians as a brutal assault on their culture, traditions and identity.'

By introducing a European-style legal system and effectively preventing clerics from becoming judges, Reza denied the mullahs their traditional livelihoods. (By 1931, their judicial functions were restricted to matters of marriage, divorce and the appointment of trustees and guardians.) By emancipating women, introducing compulsory surnames and laying down the equality of Muslims and non-Muslim minorities like the Armenians and Jews, he undermined the hierarchies that the clergy upheld. By imposing state control over large religious foundations, he intruded on their revenues.

In his memoir, Ayatollah Hossein-Ali Montazeri, who is now Iran's senior theologian – recalls Reza's banning of turbans and gowns for all mullahs except a minority, who were 'accredited' to the government. 'At that time,' he writes, 'I wasn't wearing a turban, but those who wore them lived in fear and terror.' Reza's police, Montazeri says, would burst into the sem-inarians' cells and rummage through their cupboards for the offending garments. The Shah's reforms meant that the mullahs' incomes dropped dramatically. Montazeri remembers an ayatollah who became so poor, he relied for dinner on traders in the bazaar. 'They would invite him home, for broth.'

As the mullahs had feared, Reza's inspiration in many things turned out to be Ataturk. Their modest origins and shared aspirations make them naturally comparable. Both inherited nations in chaos, monarchies being undermined by parliamentary democracy. Both desired independence from foreigners. Both were actuated by a sense of oriental inadequacy and shame. Without caring to know much about it, Reza and Ataturk reviled Arab culture and mistrusted Islam – which they regarded as a guise for Arab culture. In order to progress and gain freedom from the West, they believed that their respective countries should become, paradoxically, more Western.

Ataturk was a more convincing Europhile than Reza. Ataturk's plus fours, monogrammed cigarettes and Viennese waltz would have done credit to Edward VIII. Reza, by contrast, seemed comfortable only in military uni-form. He ordered a splendid four-poster, but the mattress was too soft so he slept on the floor. Unlike Ataturk, who knew good French and passable German, Reza never learned a foreign language, not even Russian. His Persian was simple, often painfully so. Even at his coronation, Vita Sackville-West wrote, Reza looked 'what he was, a Cossack trooper'.

Reza reportedly made this pathetic defence of his Westernizing reforms: 'All I am trying to do is for us to look like [the Europeans] so they would not laugh at us.' They did laugh. In 1935, the British minister at the legation in Tehran reported the Shah's interest in a Western invention:

Water closets are the great excitement here . . . the imminent advent of the Belgian special ambr [ambassador] and the Crown Prince and Princess of Sweden has aroused HM to a sense of the inadequacy of the Persian privy. The *sous-chef de protocole* was suddenly summoned the other day and ordered to see to it that water closets in the European style should be installed wherever the Swedish royal party were likely to descend in the course of their journey from the frontier to the capital. Next day a lorry left Tehran heavily loaded with all the sanitary fittings which the meagre resources of Tehran could furnish.

In the same year, Reza ordered all men to wear brimmed, European-style hats. (Ataturk had issued a similar edict more than a decade earlier.) Crowds gathered in Mashhad, at the shrine of the Imam Reza, the eighth Imam and the elder brother of the Innocent One, to protest. When they refused to disperse, Reza's troops mounted machine guns on the roofs overlooking the courtyard and opened fire, killing more than one hundred people. Troops who refused to fire were themselves shot.

The following year, Reza banned the women's covering, the chador. Henceforth, women were to go out bareheaded or not at all. It is hard to overstate the grossness of this edict. (Even Ataturk didn't try it.) In traditional Iranian society, the chador had been a shell; its protective anonymity had allowed women to venture freely out of doors, to shop and visit. Now, the police were instructed to rip all head coverings off women in the street. The wives of government officials had to make bareheaded appearances at ghastly lunch parties. 'The effect,' says one Iranian historian, 'was as if in 1936, European women had suddenly been ordered to go out topless into the street.' From then, until Reza was unseated in 1941, millions of Iranian women didn't set foot outside their houses.

Reza forbad the printing of Turkic languages spoken by millions of Iranians, and tried to expunge borrowed Arabic words from a new, 'purer' Persian. He forced foreigners to replace 'Persia' with 'Iran', the name used by Iranians since time immemorial, and an echo of their spurious Aryan purity. (He ignored non-Aryan minorities like the Azeris, Arabs and Turkmens.) He glorified Iran's pre-Islamic past. (Ataturk committed similar follies, briefly going as far as to introduce the Turks as the progenitors of all mankind.) Under Reza, the parapets of public buildings were crowned with stepped, pyramidal crenellations. They were meant to remind people of the glories of pre-Islamic, pre-Arab Persepolis, which abounds in these Lego-like excrescences.

Reza is remembered for murdering opponents and allies alike, for starving Iran's already emaciated democracy, for a vast, destructive paranoia. He was the kind of man who could kill a close associate the day after they had enjoyed a jolly game of cards. Peasant made good, he rarely resisted appropriating, at a nominal price, any tempting agricultural land he happened to pass. (After he abdicated, it was discovered that he owned 10 per cent of Iran's farmland – the 10 most fertile per cent.)

And yet, even now, you hear people say: 'What we need is another Reza Shah.' Iran would have changed, even if Reza had not been, but he remains synonymous with an extraordinary transformation. He turned Iran from a disintegrating empire into a nation state. He started industrialization, improved health care and education, and founded Iran's first university – to which women and men were admitted on equal terms. He answered admirably that fatalistic Persian desire to rally around a 'strong man'. (What, after all, was Khomeini, if not a strong man?)

In the final analysis, though, he suffers by comparison with Ataturk. Although he behaved like a sultan, Ataturk laid the foundations of a democracy. Reza left behind a stubbornly oriental despotism.

He has been accused of being the foreigners' lackey. This is absurd; Reza loathed his country's past subservience, but he didn't have Ataturk's genius for foreign affairs, and didn't listen to wise advice. In 1933, he extended the term of a lucrative oil concession that had been granted to the British-owned Anglo-Persian Oil Company. To compensate, he flirted with Hitler's Germany before and during World War II. (Turkey, by contrast, even after the death of its founding father, maintained Ataturk's scrupulous neutrality.)

Reza's zigzags cost him dear. In 1941, when he refused to neutralize Nazi influence in Iran, Britain and the Soviet Union invaded. Reza abdicated, in favour of his son, Muhammad Reza, and went into exile. He died, three years later, in Johannesburg.

I recently received a letter from an American lady who had travelled around Iran in the 1930s. She recalled some restrictions that were placed on Western travellers. 'One was that I could not take pictures of camels (I had no camera), nor was I to write about camels (camel trains were everywhere), because Persia was becoming a modern country. People in towns had to wear modern jackets, and other such items of clothing – to be modern, always looked uncomfortable.'

❖   ❖   ❖

Both are in the power of the church, the spiritual sword and the material. But the latter is to be used for the church, the former by her; the former by the priest, the latter by kings and captains but at the will and by the permission of the priest. The one sword, therefore, should be under the other, and temporal authority subject to spiritual ... If, therefore, the earthly power err, it shall be judged by the spiritual power ... but if the spiritual power err, it can only be judged by God, not man ... For this authority, though given to a man and exercised by a man, is not human, but rather divine ... Furthermore, we state, define and pronounce that it is altogether necessary for every human creature to be subject to the Roman pontiff.

Pope Boniface VIII, 1296

Mr Zarif remembers, as a child of about eight, being taken by his father to Mashhad, to the shrine of the Imam Reza. He was astonished by the size of the shrine complex, which is the most splendid in Shi'a Islam, and by the bazaars full of gold, and the grains of saffron, big and red, for which Mashhad is famous.

One day, when Mr Zarif and his father were in the shrine, a young man got to his feet and started to preach. The preacher was handsome and had a new beard. He wore the black turban of a *seyyed*, and a black diaphanous gown on top of a tan frock coat made of camel's wool. Mr Zarif cannot recall the subject of the sermon, only that the people were rapt. The preacher spoke classical Persian, fortified with Arabic verses from the Qoran. After the sermon, the *seyyed* walked sedately from the shrine, fondling a rosary whose beads were made of agate.

After that, Mr Zarif liked to practise at being a mullah. Back in Isfahan, there was no cleric's gown to hand so he improvised with his mother's chador. One harvest-time he climbed to the top of a hay bale and 'preached' to his friends. Whenever a mullah came to visit his father, Mr Zarif would approach him shyly and sit nearby.

In normal circumstances, this childhood infatuation would have faded and Mr Zarif, whose parents were middle class and had material aspirations, would have been pushed towards a profession; he would have ended up a doctor or an engineer – even a bureaucrat, like his dad. Instead, during the months and years that preceded the Revolution, the mullahs stirred. Some of them disappeared for months, into the Shah's jails. They were reviled by the regime and the people's esteem for them rose. They came to be seen as men of action, with frayed cuffs and an incontestable integrity. They made the Shah seem vacillating, like a shrill boy.

The people of Isfahan liked Montazeri. Although he was from Najafabad, which was several hours away, and spoke with a Najafabadi accent, he'd studied in Isfahan before joining the seminary in Qom. He was well remembered – not least for being a brilliant student, capable of recalling, word for word, lectures that had been pronounced weeks before. He had chubby lips and a flowery nose; his oratory was shambolic. He represented to many a new breed of activist mullah; these mullahs were responding to the quietists who had dominated Qom for the middle part of the twentieth century, who had kept their distance from politics. Montazeri's devotion to the people, and his willingness to expose himself to danger for them, was inextricable from his devotion to God. He lived simply. Not for him the buttery, pistachio-strewn rice favoured by the sycophants at court; he took bread and cheese, with scraps of garden cress.

Montazeri was the intermediary between Khomeini, his former teacher, and the people. While Khomeini was in exile in Najaf and then in Paris, Montazeri was one of the people who were responsible for collecting the tithes and religious taxes that were submitted by Khomeini's followers. From these taxes, Montazeri paid a stipend to the seminarians whom Khomeini was supporting. Montazeri suffered jail and exile; he endured physical and mental depredation, and the knowledge – manipulated against him by the Shah's regime – that his son was being severely tortured.

Gradually, in Mr Zarif's mind, the memory of the aristocratic preacher in Mashhad faded. The Imam and Montazeri took his place.

The return of the Imam and the vast crowds that thronged Tehran – these were the triumphs of the activists' Islam. Those who hadn't entered the Shah's anthill of preferment would be smiled on by the government of God. A few weeks later, Iranians were asked, under a specially expanded franchise, if they would like Iran to become an 'Islamic Republic'. Ninety-eight per cent – of an exceptionally high turnout – replied that they would.

But how was Iran to implement the rule of God without resorting to the political vocabulary of the West? Shi'a Islam has a temporal sovereign, the occulted twelfth Imam. But nowhere does Shi'a Islam clearly explain how humans should rule themselves in the meantime. As long as he is occulted, the twelfth Imam can hardly be the arbiter of day-to-day affairs.

In Paris, the Imam had said that 'an Islamic state is a democratic state in the real sense of the term. There is complete freedom for all religious minorities, and everyone can express his beliefs.' On the strength of these and other comments made by the Imam, many people assumed that there would be few differences between Iran's new constitution and a Western, democratic one.

It's possible that the Imam was speaking sincerely in Paris and that he subsequently changed his mind. (In Paris, he approved a draft constitution that restricted the role of senior mullahs to membership of a council that could declare legislation incompatible with Islamic law, at the request of government officials.) Some believe that the Imam was exercising what Shi'as call prudential dissimulation – the right to lie about one's beliefs as long as the lie is in the wider interests of the faith. Others think that Khomeini had little idea, when he was in Paris, of what kind of government he wanted to install. Soon after he returned to Iran, it became clear that his intention was not to transplant Western democracy to an Eastern host, but to implement pure Islamic rule.

For the origins of this, it's necessary to go back to the two centuries that elapsed between the fall of the Safavids and the rise of Reza Shah. During this period of administrative feebleness, Shi'a Islam reduced its links with the state. It diversified. Among themselves, the mullahs debated the responsibilities that God expected them to shoulder.

The outcome was a self-conferred expansion of their mandate. Advanced *ejtehad* – the exercise of rational judgement in the application of Islamic ordinances – gave senior mullahs unprecedented freedom to elaborate existing laws. During the eighteenth century, furthermore, the Shi'a community was divided into imitators and imitated; every believer was obliged to select an 'object of emulation' from the top rank of *mojtahed*s. (*Mojtahed*s had yet to award themselves the honorific ayatollah, which means sign of God.) The believer would bow to the *mojtahed*'s superior knowledge – to his ability to deduce God's ordinances according to logical proof. No *mojtahed* had absolute authority. The rulings of one might contradict those of another. But, logically speaking, the *mojtahed*'s learning and experience meant that he was bound to be right more often than a non-*mojtahed*.

Some people found unacceptable the subservience that was inherent in this relationship. They sought out esoteric paths. Some, using Sufi mysticism and the intercession of holy men, contracted personal relationships with God. Others tried to unearth the buried guidance of the occulted twelfth Imam, and to make that the determinant of behaviour and policy. But, for millions, especially the illiterate and uncertain, the principle of emulation was a soothing abnegation of responsibility. As long as they did what their chosen *mojtahed* told them, they could count on being admitted to paradise.

As the nineteenth century wore on, the clerics became a force and they were famous for their recalcitrance. They loathed Fath-Ali's son, Muhammad, who reigned for fourteen years until 1848, because he patronized Sufis. They denounced Muhammad's successor, Nasser al-Din, because his forty-eight-year reign increased Iran's contacts with infidel Europe, as well as its indebtedness to Jewish banks and Christian governments.

In 1891, to protest against the auction of Iranian sovereignty, an influential *mojtahed*, Mirza Hassan Shirazi, trespassed sensationally into politics. In response to the tobacco monopoly that Nasser al-Din had granted to a British company, Shirazi commanded a boycott of tobacco products. Around the country, water pipes were laid down. (Even the Shah's wives are said to have obeyed the edict.) The monopoly was cancelled.

From then until Reza Shah's reign, the clergy's involvement in politics deepened. Senior mullahs coordinated opposition to some detested foreign loans. Two *mojtahed*s led protests that forced Muzaffar al-Din Shah to accept the inception, in 1906, of Iran's first parliament. A second group of mullahs, sniffing the air and detecting European republicanism, opposed this alien innovation. Parliament, it was possible, might enact legislation that contravened Islamic law. One of these opponents, Sheikh Fazlollah Nouri, who was later executed, favoured a pure Islamic government. He advocated an assembly whose only responsibility would be to implement Islamic law.

In around 1970, while he was in Najaf, Khomeini delivered a series of lectures, subsequently collected and published, that put flesh on the bones of Nouri's longing. It was illogical, Khomeini argued, to suppose that God desired Islamic government only so long as there was an infallible guide – the Prophet or the Imams – to oversee it. Even now, during the occultation, the government of God should be implemented. Who was better qualified, he asked, to oversee Islamic rule, than the doctors of Islamic law – the senior jurists, or *mojtahed*s?

In the middle of the nineteenth century, Sheikh Morteza Ansari, the senior *mojtahed*, had awarded himself administrative power over possessions and people. But Ansari had restricted this 'guardianship' to orphans – and to others, like the mentally ill, who were incapable of managing their affairs. Using deductive reasoning and a novel interpretation of some sayings of the Prophet and the Imams, Khomeini enlarged this principle. Accordingly, the whole Islamic community effectively became an orphan – the twelfth Imam being the absent 'parent' – and the jurists were the guardians of society. The responsibility of these jurists, Khomeini argued, was to set up an Islamic government, putting at its head either the pre-eminent jurist or a college made up of several jurists. This 'Guardianship of the Jurist' would, of course, render the institution of the monarchy superfluous.

Khomeini cited a saying of the Prophet, supposedly transmitted by the Imam Ali, as evidence that the clerics are the successors of the Prophet. The Prophet was asked who would succeed him, and he replied, 'Those who will come after me, pass on my traditions and put them into practice and teach the people after me.' Khomeini presented this as an allusion to the clerics.

At the time of the Revolution, few people had heard of the Guardianship of the Jurist. After Iranians voted for an 'Islamic Republic', clerics around Khomeini started actively to promote his idea. They are said to have manipulated elections to an Assembly of Experts, headed by Montazeri, which was mandated to draft a constitution; two-thirds of the assembly were mullahs. There was disagreement over how much weight should be given to Western institutions and concepts – such as parliament, accountability and elections. (Some argued, for instance, that the massive crowds that had greeted the Imam on his return from Paris constituted an irrevocable oath of allegiance, rendering elections meaningless.) Most were united, however, on the need to crown Iran's new edifice with the most charismatic jurist of the day: Khomeini.

Montazeri admitted that the 1979 constitution was 'not ideal . . . it is a compilation of different views put together with the central objective of producing a constitution that is in accordance with Islam'. Examining it, as long as you confine your view to the bottom layers of the pyramid of power, you get the impression that power emanates from the people. One article states that no one can deprive people of their 'God-given right' to determine their 'social destiny'. You notice features and disciplines – such as a parliament and a president that are elected by universal suffrage, and the independence of the executive, legislative and judicial branches – that make you think of Western models. If you look further up, however, you're confronted by a strange bifurcation. The tip of the pyramid is barely connected to the mass; the senior jurist – the Guide – might just as well have been transplanted from heaven.

According to the 1979 constitution, the guardianship of the community should devolve upon a 'just and pious jurist, well-informed with his times, courageous, resourceful; ... recognized and accepted by the majority of the people'. But no mechanism for gauging this popular acceptance is adumbrated, save elections to the Assembly of Experts, whose membership is heavily influenced by the Guide. Look further and you'll see that the Guide and his representatives have a veto over the whole government. In theory, the three branches are separate. In practice, the Guide can manipulate all three.

According to Montazeri, the ruling jurist was not meant to be 'above the law'. But Khomeini was not merely above the law – his word, literally, was the law. He was one of four top Shi'a divines then living in Iran; millions of people regarded him as their personal object of emulation, as well as their unrivalled political leader. In the person of Khomeini, religious and political expertise merged, surmounted by an extraordinary charisma. Some likened Khomeini to a 'philosopher-king' on the Platonic model, and noted that the clerical class he headed was not dissimilar to the disinterested elite that runs affairs in Plato's Utopia. The clergy had come a long way since Ayatollah Shirazi's ban on tobacco consumption.

According to Montazeri, Khomeini didn't envisage clerical government when he was in Paris. Soon after his return to Iran, he repaired to Qom, notionally leaving the reins of power in Bazargan's hands. But this was a fiction. Wherever you looked, there were relatively transparent – sometimes even democratic – institutions that were fronted by non-clerics. The trouble was, all of them had ill-defined clerical shadows. The government was shadowed by the Islamic Revolutionary Council (IRC), which guided policy. The IRC's membership was not formally made public, but it was weighed down with mullahs, and arrogated to itself the right to enact 'legislation'. The army was shadowed by the Revolutionary Guard – the mullahs' militia. The foreign ministry was shadowed by Khomeini's and Montazeri's mostly clerical representatives, semi-official foreign envoys who whizzed around promoting Iran's most important export: Islamic Revolution.

When the first parliament of the new regime opened, in 1980, a large minority of its members were clerics. Clerics ran the Foundation of the Oppressed, which confiscated the assets of tens of thousands of affluent Iranians who had fled abroad, as well as swallowing some 20 per cent of the assets of all private companies that existed before the Revolution. The mullahs were everywhere. They carried account ledgers, policy papers, Kalashnikovs. It was their God-given right.

✦ ✦ ✦

Mr Zarif was seventeen when he entered the seminary. He'd spent months at the front. He'd killed men, though not at close quarter, and seen people he knew being blown up. He'd watched the Islamic Republic come into being and rejoiced at the birth of the Guardianship of the Jurist. Now, it was 1982, a few weeks after Khorramshahr. For the past three years, Mr Zarif's pulse had been racing. Three years of grown-up memories seen through maturing, regressing, adolescent eyes. (Too many memories to fit inside his head? It sometimes felt that way.) He wanted to sit and read, to thrive slowly.

Mr Zarif has written almost a memoir about his time in the seminary. It's one of his many unfinished enterprises. (Mr Zarif seems to derive pleasure from striving at some work of writing, only to lay down his pen shortly before the task is completed.) I have read the nearly-memoir. It's lush and generous. When Mr Zarif went from the battlefield to the seminary, he was seeking beauty.

The beauty is not apparent, not immediately. Not in the dryness of Qom with its thudding sun and *sowan* – the parched caramel, embedded with pistachio shards, that masquerades as a local speciality. The beauty isn't obvious on a casual visit to one of the seminaries. Their plans and proportions recall Isfahan's Four Gardens Seminary, but most of them date from the second half of the twentieth century and are appallingly built. Some are made of concrete that has been clothed in brick-like slabs. Machine-made tiles affront your eyes if they alight upon a spandrel. Bits of wall flake off. Renovations are ongoing.

The seminary looks in on itself; it has no façade but a postern sunk in a wall with a kiosk for an impolite porter to nest in. The college's modest visual entertainments – tubular pines, parterres crammed with petunias – are guarded from public view. Entry is barred to anyone without good reason to be there. It's doubly barred to women. The seminary is a barren paddock for exercising the soul. (Think no more of the glance of a brazen girl, shopping with her mother, in the nearby bazaar! When you're inside a seminary, the bazaar seems as distant as Amsterdam.)

The seminary is a privilege and a burden. You're scrutinized; the populace admires the rigour of your enquiry. But how many respectable *bazaaris* would marry their daughters to a seminarian? Not many. (Of course, the traders would make an exception for the nephew of a senior ayatollah; that would make of them nearly a kinsman of God's.) Then, there's the silence, the unnerving tranquillity. It's no wonder that many young seminarians get homesick.

The law, as you must know in order to be a good Muslim, exists. It does not need to be made. It is immutable. The sources – the sayings of the Prophet and of the twelve Imams, and the example set by the Prophet and the Imams – are the mullah's raw materials. (Believers are enjoined to follow the Prophet's example as closely as they can, though this doesn't extend to his dozen-odd marriages.) Some of the law lies at the surface. (For example, the Qoranic injunction against drinking alcohol, which is written down, and anyone can understand.) But much more of the law needs extraction, like ore.

Mullahs study so that they can use their knowledge of God's will, and their God-given rationalism, to extract the law. They have tools: their knowledge of Arabic, grammar, rhetoric and logic. These help them when it comes to learning a set of principles for deriving the law from its sources. The art of derivation is called jurisprudence.

The first thing that Mr Zarif learned was Arabic literature. The Qoran isn't solely the word of God, transmitted through the Prophet; it's the trophy of Arabic literature. The sayings and traditions of the Prophet and the Imams are full of literary artifice and allusion. A thorough knowledge of them is one of the prerequisites to understanding the law.

After a while, Mr Zarif moved on to logic and rhetoric – the arts of language – and to memorizing chunks of the Qoran. You need this knowledge before you progress to the principles that govern jurisprudence. (Only then can you turn to the detail of the law.) It's acceptable for *mojtaheds* to reach different conclusions, as long as they reach them using universally accepted methods.

Mr Zarif's classes took place in the morning. After lunch, he would meet his study partner and they would discuss the text they had read in the previous class, and prepare whatever had been set for the following one. Mr Zarif and his study partner liked to spar with each other. Whether he was discussing the difference between denotation and connotation, or the superiority of an imported Buick over an Iranian-made Cadillac, Mr Zarif discussed seriously, with a respect for the rules. He learned to dislike sloppy argument, the corollary of sloppy thought.

The important texts have been densely written. You can't study them in a hurry. Mr Zarif advises: 'Kneel before the book in front of you with a calm as substantial as the four or five centuries that have elapsed since the book was written. You have to release your entire being, to make yourself believe that nothing exists in the world except for understanding the book in front of you. This requires relaxed nerves and an utter absence of excitement. This is the treasure of the seminarian, more valuable than a lifetime spent in pointless activity.'

The aim of all the effort is eventually to be able to answer the people's questions. That's what mullahs are for. You generally get to this point after about six years' study. One day, your teacher asks you if you've thought of putting on a gown and turban. (Until now, you've been wearing normal clothes.) It's a compliment, for the regalia are an announcement to the world, an expression of confidence in your abilities on the part of the profession. You blush and say that you're not ready. This is *ta'aruf*, of course. He'll ask you again and you can relent under pressure.

Some mullahs aren't interested in the people or their questions; they rarely step out of the seminary. Some want to teach. Others want a quiet life, without stress and bother. If they marry, their cells may be passed on to a son, with familiar smells and unwashed tea things. (The Shi'a Imamate, remember, is a holy family, founded by the Prophet; Iranians have an innate respect for the hereditary principle.)

Nowadays, the seminaries are run more like a modern university; set books have to be read within a certain time and there's been an increase in the number of exams you have to get through. In Mr Zarif's day, life unfolded at its own pace. The teachers might advise you to try and read a set text over the course of the coming year, but some students might finish it in six months and others in two years. There was no particular stigma attached to those who read slowly; it didn't affect the stipend they received.

Mr Zarif found the beauty, in the meditative life and its distance from the War – in the texts. Argument and the superstructure of scholarly disputations were part of that beauty. But Mr Zarif soon discovered that integrity and virility in scholarship meant little without certain moral qualities that Islam, according to his understanding, requires from its adherents. As you would expect from an institution that is governed by a strict hierarchy, the seminary is full of pride. Mr Zarif came to dislike that pride, and to seek its prickers.

He remembers an itinerant, an impoverished old man who squatted in a derelict house and made ends meet singing laments in the street about the Imam Hossein. They called him the *Seyyed* because he wore an old black turban – it was an ironic name, for no one believed he actually was a descendant of the Prophet. The *Seyyed* was tall and stooped. Mr Zarif enjoyed bumping into him; the *Seyyed* was sharp in repartee. But Mr Zarif would always defer his invitations to visit. The *Seyyed* was known to frequent prostitutes. They would sit at his knee, it was rumoured, and weep for their lost grace. Some suggested that the *Seyyed*'s interest in them was more than pastoral.

One day, Mr Zarif asked one of his teachers to recommend a mullah who could give him extra tuition on a text he was having difficulty with. The teacher said: 'Do you know the singing *seyyed* who wanders around the streets? Go and see him.'

'The *Seyyed*?' Mr Zarif exclaimed. 'He's a bum!' But the teacher's words had provoked his curiosity. A few evenings later, he knocked at the *Seyyed*'s door. The *Seyyed* asked him in, and Mr Zarif was surprised to see that the house, although bare and mean, was spotless. The *Seyyed* invited Mr Zarif to sit on the floorboards – he didn't seem to own a carpet or rug – and started to recite for his visitor's entertainment the different miaows of the thirty-odd cats that were living in the upstairs part of the house. Mr Zarif interrupted him and told him the purpose of his visit.

An hour later, when he rose to leave, Mr Zarif had a different idea of the *Seyyed*. The *Seyyed* had become a brilliant teacher and a tricky text had started to make sense. On his way out, Mr Zarif's eye was caught by a photograph of a young mullah standing next to Ayatollah Khoi, who had been one of the great clerics of Najaf. Mr Zarif turned to the *Seyyed* and exclaimed, 'But that's you! You were a pupil of Ayatollah Khoi's!' Mr Zarif stopped himself, and they both knew what he wanted to ask: 'You, who had such a bright future! What went wrong?'

As the *Seyyed* opened the front door for Mr Zarif, he said, 'If you're seeking knowledge, you must fly from position and power. Fly!'

For the next few months, Mr Zarif came once a week to the *Seyyed*, until he fully understood the text. The *Seyyed* refused to take money or gifts, so Mr Zarif sought him out during the day, when the *Seyyed* was someone else, and slipped him pennies for his laments. After their final class, Mr Zarif hesitantly broached the subject that had troubled him since he first set foot in the *Seyyed*'s house. He said: 'I've been told that you spend time with women of ill repute. Why do you do this?'

'I see these women for their purity.'

Mr Zarif said: 'You mean they benefit from your status as a cleric?'

'No, I mean that I benefit from their purity.'

Mr Zarif frowned. 'Where is this purity?'

'It's in their belief that there exists nothing so insignificant as them on the face of the earth. They've been so crushed beneath the weight of sin, there remains no trace of pride in them. And when they say they're sinners and worth no more than the cur which sits at my door, I know they're not like the charlatans we see around us.'

The *Seyyed* put his hand on Mr Zarif's shoulder. 'They're a step ahead of us. They know they're nothing before God, whereas we persist in thinking that we're something.'

❁ ❁ ❁

In the seminary, you're closer to God than ordinary people, but that doesn't make you closer to heaven. There are three threats to the tranquillity of the seminary. Philosophy is one.

Some of the philosophers have performed great services for Islam. Take Abu Ali Ibn Sina, whom Westerners know as Avicenna – the man who, at the beginning of the eleventh century, did much to incorporate Aristotelian logic into the mainstream of Islamic thought. Avicenna adumbrated a 'proof' of the existence of God that influences thinkers and theologians to this day.

In his 'proof', Avicenna posited a necessarily existent essence, God, as sustaining the realized existence of all other essences, which are only possibly existent. He and other Muslim thinkers furnished reasons why people should obey God's law. (Reasons that are sturdier than habit, more honourable than inheritance.) Philosopher-clerics argue that a mullah without philosophy is like a parrot, or a whistle in someone else's mouth – making the right noise but unable to think.

But the anti-philosophers are numerous. They disapprove of philosophy because they see its power to destroy. (They regard it racially – a Hellenic trick that has on occasion proved useful for splitting the Islamic community.) If the Prophet and the Imams were linked directly to God, what use is there for confirmation of what we know, on far better authority than human reasoning, to be right? Philosophy, they suggest, places the onus on Islam, which was revealed and cannot be questioned, to prove itself. At best, this is impertinence. At worst, it's blasphemy.

Just as philosophy can prove, its opponents argue, it can also disprove and induce doubts. There's nothing more threatening to the seminary than a student with doubts. His frailty spreads like ivy, strangling the certainties of others. He irresponsibly popularizes Socrates' declaration, 'One thing only I know, and that is that I know nothing.' The seminary – the management, teachers, and students – must be vigilant.

In Avicenna's time and beyond, some traditionalists regarded systematic methods of analysing religious language as threatening and alien innovations, which would inevitably bring dangerous thoughts in their train. Eventually, the art of disputation, which is partly the art of systematic reasoning, became prevalent in Islamic places of learning. But prejudice against philosophers remained. When, in 1191, Saladdin, the warrior-king who had conquered the western and southern Mediterranean in the name of Islamic orthodoxy, ordered the execution of a controversial Persian thinker, the condemned man was traduced as someone 'who believed in the system of the philosophers among the ancients'.

Ayatollah Montazeri recalls that a mistrust of philosophy prevailed in

Qom when he was a young teacher. The popularity of two famous classes of philosophy unnerved Ayatollah Burujerdi, Qom's senior cleric at the time. He threatened to cut the stipends of students who attended them. Burujerdi and others regarded philosophy as a danger, better restricted to a few seminarians whose beliefs were rock solid. The last thing they wanted was that a class of philosophy become a distraction from the main business of the law. There was a story going around, Montazeri recalls, of a student emerging from studying a text that teaches the oneness of the intellectual subject and object, and proclaiming, 'If there is one existence, that means I'm a piece of God.'

❖ ❖ ❖

The second danger is the night. There are no classes at night, no formal prayers. Some seminarians understand the daylight only in opposition to nights that fluctuate, becoming taut and limber apparently of their own volition. When the mystical seminarian stands on the holy threshold, time stops and races until the morning call to prayer flings its dawn across the sky.

They are the seminarians who don't have the patience to wait, who demand an immediate audience with God. They're in pursuit of some knowledge that can't be taught. They perform small abstentions. You might hear, for instance, of Asghar, a young seminarian, fasting outside the month of Ramadan, or denying himself sleep one night every week. The same Asghar – not that you'd know it, for he conceals his inclination – might spend days on end repeating to himself an invocation.

Asghar raises his hand to make a point during a class, has a disputation with his study partner. He gets angry with a letter from his mother, who advises him to marry. All the time, unknown to anyone – sometimes, he feels, unknown even to himself – he's repeating: 'There is no power and no power save in God!' Again and again: until the phrase has entered the tissue of his life, until he's woken up by it at night.

'There is no power and no power save in God!'

Gradually, the words lose their echo; they cease to be words, become an essence – dissolving in his faith. There is nothing that is not a manifestation of God. Everything is God. There is no power and no power save God. Whoever has known himself has known God too and knows all that the Prophets knew. As the poet Rumi, whom we call Our Teacher, wrote:

> I speak of plural souls in name alone –
> One soul becomes one hundred in their frames;
> Just as God's single sun in heaven
> Shines on earth and lights a hundred walls
> But all these beams of light return to one
> If you remove the walls that block the sun
> The walls of houses do not stand forever
> And believers, then, will be as but one soul.

Inside his cell, whose light is never extinguished, Asghar is putting words together inside his head. There's a border between this and a remoter passion – the possession and ecstasy of heretics. Where does the border lie?

Take the old man – there are millions like him – who spends many of his waking hours counting with his fingers the ninety-nine beads of a rosary. He's invoking the ninety-nine different appellations for God in the Qoran. What is the Islamic arch, strung in pendulous repetition across a river, but an invocation of God?

Remember excess. Remember Hallaj. He was the divine of tenth-century Baghdad. His pursuit of God induced in him a mystical inebriation, and he revealed his relationship with God. Submersion in the divine had turned Hallaj's head. Before his followers, he proclaimed: 'I am the Truth.' For which blasphemy, naturally, he was executed.

What frightens the authorities is that Asghar, having discovered a new channel to God – direct, intuitive – may come to consider gnosis superior to the law. Asghar will become deluded, arrogant. (He may start to regard daily observance as unworthy of his God-like wisdom, and abandon it.) He's putting words together, extemporizing verses that etherize upon utterance. He's waiting to see the light that has been promised to the mystics, a soft light that imbues everything – radiating from inside him, from the core of an apple, in the wing of the fly that's buzzing in the window. He chants softly, pricks his skin with knives, pomegranate blood. He feels no pain.

To talk about lust, Mr Zarif and I have repaired to a teahouse in Qom. It wouldn't do to have such a conversation in the presence of Ali – who is, after all, the fruit of his parents' lust. I remark that the teahouse is an odd choice of location. Teahouses are sometimes coarse and dirty but rarely tolerate prurience. This teahouse is almost empty, save for a couple of dank-looking addicts convalescing with tea on a day-bed in the corner, and three young men, seminarians in civvies by the looks of it, conversing over a water pipe. It's quiet, but Mr Zarif reassures me that our conversation will be drowned out by the chatter of some finches that are imprisoned in a cage near our table.

I've written down some words in Persian on a piece of paper. Masturbation; wet dream (the Persian–English dictionary suggests nocturnal pollution); erection; sodomy. I intend to deploy these words during our conversation. I hope I don't mispronounce my key words and have to be corrected by Mr Zarif. Our waiter shuffles over with tea and a water pipe and squints at us suspiciously. When he's shuffled off, we begin.

On the one hand, in Islam we have miniatures of Shahs and Sultans: hunting, eating meat, practising at falconry, being served wine by rose-lipped pages and hirsute princesses. We know, also, that Islam is not a renouncing religion. You don't get mullahs following the example of monks and nuns, contracting pious matrimony with the Church. (In Islam, because religious power has never been centralized, there's no 'Church' – no manifest divine authority that has to be courted, fed and nursed through its dotage.) Sex in Islam isn't shied away from, either. Islam wants you to have sex properly, in accordance with the wishes of God.

But when Mr Zarif attended his first morals class in the seminary, they told him that lust – besides covetousness and anger – was the work of Satan. They told him: 'If you want to approach God, you must combat your desire.' That meant no rubbing, no fantasies. The trouble was, the people who said this were themselves married, and their children's existence attested to their carnality. 'The authorities,' Mr Zarif says, 'knew they couldn't deny lust, so they tried to camouflage their ambivalence by speaking about it obliquely. If you pressed them, they'd say that there are two kinds of lust: the healthy kind and the unhealthy kind.'

Mr Zarif smokes. He hands the snake to me and says: 'There are seminarians who deny lust. They try to starve their sexual desire; they don't acknowledge it, hope that it will die. They convince themselves that study and prayer will answer their needs. They end up nervous and distracted. Some of them end up having homosexual affairs.' The finches are behaving riotously, which is just as well, since we've strayed from standard teahouse topics.

'There's another group in the seminary,' says Mr Zarif. 'They're the sort of people who listen to the morals teacher with a mocking smile. They're the ones who contract temporary marriages. They keep that hidden from all but their friends, of course.'

'Why do they have to hide it? Temporary marriage is legal.'

'Yes, but it's not something the seminary approves of.'

'And what if they come out of the legal framework? What if they have unmarried girlfriends; that would be legally and religiously impermissible.'

Mr Zarif's face becomes blank and hard. 'Then, they stop being mischievous. They become unhealthy, and we hate them.' He says this without effort. The words require no elaboration, and I'm reminded that Mr Zarif, for all the generosity and subtlety of his friendship, is rigid in his belief and fanatical in its defence. The roisterers – they're not bending the rules in any forgivable sort of way. They're flaunting their deviancy, making fun of God. They set themselves up to be the spokesmen of Islam and, from a pedestal of privilege and trust, commit their squalid betrayal.

Mr Zarif says, 'My friends and I, Mr de Bellaigue, were part of neither of these groups. We fortified ourselves by creating ideals, and trying to reach them.'

He doesn't deny the struggle. His brightest memories are sensual. He remembers half a dozen Baluch students – boys from the boiling hot Pakistani border – gathered in the seminary courtyard one winter morning, after a rare fall of snow. The Baluchs had never seen snow; they were holding it up in their hands and letting it melt and fall through their fingers.

At night, Mr Zarif tells me, he and his friends would gather in a cell and talk about poetry and tell each other stories about things that had happened to them before they joined the seminary. When they got hungry, they'd have some rice or fried potatoes and eggs. Later, after the dawn call to prayer, they'd return to their own cells and go to sleep.

'Did you talk about sex?'

'There was talk – we certainly didn't pretend that it didn't exist. But we'd never use the sort of words for sexual communion that you'd hear in the bazaar. One of the boys might have news of a woman who was seeking a temporary marriage, but it was understood that none of us would take up the offer.' Maintaining purity was a collective effort. If the ideal were betrayed, everyone would be contaminated.

When he was in the seminary, Mr Zarif admired a mullah who, besides the standard curriculum, knew a great deal about wine, dancing, obesity and gambling. A doctor of the law who does not have this knowledge, he argued, cannot administer rulings on them. His mastery was purely vicarious.

'What about dirty jokes?' Once, before an internal flight, I'd misplaced a verb when asking the stewardess when we would take off, and had asked her when we would have it off. Mr Zarif had enjoyed that story.

'There were jokes, but they were refined.' A superior sort of smut, whose exemplar is a book, *Zahr-al-Rabih*, which was written by seminary teachers over the course of several generations. *Zahr-al-Rabih* contains dirty stories and jokes, besides more esoteric titillation. For instance, there's the story of a man who desires a young boy and is rejected. The man delivers a long and learned peroration, drawn from jurisprudence and philosophy, in an attempt to prove to the boy that he should render himself.

Mr Zarif, I think I understand across the twittering of the finches, is not a masturbator. What, I ask, of nocturnal pollution in the seminary?

'It could be a great relief, but it upset the rhythm of my mind. After it happened, and my thoughts were taken up with the memories of it, it might take a day or two before I could start working effectively again.'

'Didn't you feel envy towards those who were in love with girls?'

'I can't remember any time in my life that I wasn't in love with a girl. From my childhood onwards, I was always in love, but the girls in question never knew.' (The only one who found out became his wife.)

'So you felt no resentment that the seminarians were acting on their love, that it had material meaning for them?'

'I reached a point where I felt that I didn't have anything less than them; on the contrary, I had something more than them.'

As we sit, I find myself digging into my own past and unearthing memories – of not having a girlfriend, or not having the right kind of girlfriend, or wanting someone else's girlfriend. Would it have been nicer not to have wanted, or to have wanted simply because it was a delightful and stimulating thing, this wanting, without expecting to receive?

The finches have shut up. From our neighbours' table there's the sound of bubbles rising through water. Mr Zarif gets up to go to the lavatory. As he's putting his feet into his shoes, whose backs have been squashed like slippers, a thought strikes him. He looks back at me, and says softly: 'What's alcohol like? You know, wine.'

'You've never drunk it?'

'No.'

'It's like a plate that everyone sits around and eats from. You should do it with other people. It turns strangers into comrades.'

'You can't do that with a bottle of Pepsi?'

'Wine has a chemical effect, it makes you happy.'

'Tell me, frankly; do you think a society is better off if it is teetotal?'

It's true that everything that seems attractive about alcohol, the levity and solace – they exist independently of it. These things aren't created by alcohol; in fact, their apparition under its influence may be false. I admit these things reluctantly and immediately regret having done so. I'd like to jab a needle into Mr Zarif's smug, certain shell.

When he's gone, I look around the teahouse. It's not a very classy place. There's a family room upstairs, where men take their wives and you can have kebabs wrapped in bread like a mattress. There's a cat somewhere around; I see its tail disappearing under a day-bed. Vileness exists so that beauty can be delineated – it's the shadow that defines virtue. But Iran has beauty and ugliness in the same elusive hue.

Someone, I notice, came in while Mr Zarif and I were conversing. The newcomer is talking to the waiter: '. . . and his chest starts to tighten, God forgive him, and his son tells him, come on dad and have a lie down on the new couch which they'd been saving up for, it has a reclining mechanism which should help for his dodgy back, you know. The son says would you like some tea, and he says, yes, he would, so while the son is off in the kitchen the dad dies right there on the new couch, and when I went to the poor bastard's burial I looked up to the sky and said please God bring me a nice comfortable death soon, with my boy bringing me tea, and the couch will be fine just fine . . .'

Mr Zarif is coming back from the lavatory – where he's certainly widdled according to the prophet's example, squatting so the piss doesn't splash on his trousers – and he's carrying a second water pipe. I put the first one, which is dead, on the next-door table. Mr Zarif sucks at the new one so the tobacco will take.

I long to say: 'Isn't doubt the only worthy companion in life?' But he's built his sturdy little family on an absence of existential doubt. He won't abandon it. His bristling intellect will defend it.

I say: 'What about Muhammad, the mystic in the seminary? In your memoir, you talk about his friendship with one of the other seminarians. I'd like to know about this.'

It's 1983. A slim village boy arrives in the seminary, and Muhammad falls in love. The village boy is the light he's been waiting for. In his face, God's light has its source. If Muhammad views things in the light that he casts, they become celestial. Can't anyone else see? Muhammad and the village boy become inseparable. Muhammad is a moth around him. They become cellmates. They inhabit a world of their own, rarely acknowledging other people. They talk to each other in whispers, and no one knows what they're saying.

I think of Our Teacher, Rumi, in 1244. That's the year Rumi was first dazzled by Shams, the holy man of Tabriz – not for nothing that Shams means sun in Arabic. Rumi spent months in seclusion with Shams and came out a different man. Before, he'd been a respected but conventional religious scholar, spending his time praying, sermonizing and studying the law. Now he was repudiating much that is exogenous, finding God within.

'Knowledge that doesn't take you beyond yourself,' Shams told him, 'is far worse than ignorance.' Under Shams' guidance, Rumi started to whirl ecstatically, accompanied by music or an invocation, and to develop his skills as a mystic. He became a poet.

> You are my sky, I am the earth, dumbfounded:
> What things you sprout from my heart each moment!
> I'm parched earth, rain down on me drops of grace,
> For your water makes the earth grow rosy.
> What does the earth know what you sow in it?
> You made it pregnant, you know what it bears
> Every atom pregnant with your mysteries –
> You make it writhe a while in pangs of birth.

Mr Zarif says, 'I've often asked myself: if that village boy hadn't been beautiful, effeminate, if his body hadn't been willowy, would Muhammad have fallen for him? Maybe lust can be an expression, or a tool, of Platonic love. You see, I'm sure that, even if Muhammad and the village boy had been left to their own devices, even if they'd enjoyed complete freedom away from the seminary, they wouldn't have had physical relations.'

'My belly's pregnant with you,' said Rumi to Shams. 'When will I see a child born by your fortune?' Rumi's disciples hated Shams for monopolizing their sheikh. This hostility drove Shams away but he returned. He married a girl who had been brought up in Rumi's household. Then, in 1247, he disappeared for good. Rumours abounded of his murder at the hands of Rumi's disciples. Rumi's mourning took years.

> I sleep and wake in love's afflictions,
> My heart turns on the spit of passion's fire,
> If you abandoned me to refine me best,
> You are wise and I'm without you unrefined.

Inside the seminary, there's been talk of the cell light, which stays on all night – for what reason? There's talk of Muhammad's compilations; words are coalescing, becoming poems. One day, the seminary supervisor orders the village boy to move out of Muhammad's cell. Mr Zarif's memoir describes Muhammad's reaction to the catastrophe:

'Muhammad grew agitated; he couldn't accept the order. He paced up and down his cell, and growled under his breath like a lioness. He was a woman in labour, who can't bear to sit and take the pain, but whose every movement increases it; each moment increased his mental agitation . . .

'Muhammad sat down and composed a lengthy poem. About the supervisor who had oppressed him. About his freedom to choose, which had been trodden underfoot. He declaimed this poem, and . . . it was as if unending gulps of divine libation had been transformed into anger, swilling in his mouth, became boiling and piquant . . .

'The grossest words were arranged in the most delightful harmony, transporting the listeners to heights of delight and astonishment. In his pride, Muhammad had several copies made and distributed around the seminary. With his own trembling hands, he presented one to the seminary supervisor and one to the director. The director was a learned man . . . just when we expected to hear that Muhammad had been expelled from the seminary, news spread that the director had torn a strip off Muhammad and rewarded him with three thousand *tomans* for his beautiful poem . . .'

It's getting dark outside. Mr Zarif calls for the bill. The waiter appears. He counts the money, and Mr Zarif says: 'Soon after that, Muhammad got married, and the relationship ended.'

CHAPTER FIVE
Lovers

I was back in Isfahan, drawn to Hossein Kharrazi because he's dead. He'd been snapped off while he was still handsome, in his prime, an Isfahani Alexander. His picture was everywhere for the popular edification – on the walls of the city, in official books that had names like *Lovers' Fervour*. Hossein joking with some Revolutionary Guard frogmen – one of them strikingly prepubescent – before an operation; Hossein breakfasting frugally flanked by Mohsen Rezai, the head of the Revolutionary Guard, and Rahim Safavi, his old pal from Kurdistan days; Hossein and his lads grinning over the lip of a trench, someone's hand hovering playfully over his head, like Piero's dove or a halo.

In many of the pictures there's no right arm – there's a background or an empty, hanging sleeve. To begin with, this absence makes you uneasy. But the more you get to know Hossein the less you get this feeling. After a while, there's no difference between Hossein before and Hossein after. Someone swiped his arm; they might just as easily have swiped his car radio. Hossein carried on. He'd have carried on no matter what bit of him fell off . . . 'Hossein, you can't lead the men into battle without your head . . .'

I went to see a veteran called Amini. He owned a kiosk near one of Isfahan's big hospitals, selling synthetic fruit juices and tea in plastic cups. After the War, Amini went into business and was unlucky. (Someone, his partner I think, swindled him.) The kiosk was the dregs of his empire. I hung around for an hour and a half and bought two mango juices, one of which I gave to his employee. Later, Amini phoned my hotel to apologize for standing me up and to set another date but his voice told me he was more suspicious than sorry.

You get used to suspicion. It's partly inherited. Iranians resent their recent past with its foreign capitulations. (Two centuries of semi-colonization sometimes seem worse than unambiguous colonization; at least the unambiguously colonized got railways and sewers and unambiguous independence.) As a consolation, many Iranians present the efforts of foreigners to subjugate them as a compliment, albeit backhanded – proof of what an important and desirable place Iran is. When I'm feeling sour, I reply that, on the contrary, Britain was only interested in Iran because it lay on the route to India.

I'm doubly or triply suspect. I speak their language, I ask questions for a living, I'm British. Spy! Troublemaker! Agent! I try and smile phlegmatically. This isn't easy; those who 'out' me as an operative of MI6 generally think they are first to hit upon this theory.

Conspiracies are an immensely enjoyable diversion and they can be catching. The longer I live in Iran, the more I involuntarily entertain the widespread and insane (surely!) idea that the Islamic Revolution was wrought by the British, jealous at having lost their dominant position in world affairs to America. Sometimes, I find it far-fetched that I should have married my wife because of love, not strategic interest, that I studied Persian at university for reasons other than national calculation, that I seek my own advancement and happiness, not a reprise of Britain's global domination.

As I walked from my hotel to meet Amini, my heart was heavy with the thought of having to overturn this mistrust. Coming down the road that runs parallel to the river – the two are separated by a lawn and trees – I saw a bus parked by the kerb. According to the bunting on the back window, the bus had come from Ahwaz, almost three hundred kilometres to the south-west; it was part of a convoy that had taken pilgrims from Khuzistan to Khomeini's tomb on the outskirts of Tehran. (It must have stopped for the night on the way home.) As I approached, I saw a line of feet protruding from the lawn at the side of the road. I couldn't see the bodies they belonged to; they were obscured by the shadow of the trees. As I got closer, I heard a hum coming from the vicinity of the feet. There was a smell of sweat and something else, an organic process like meat decomposing in the gaps between teeth.

There were twenty sleepers, male, and they had five-day stubble, and long-sleeved shirts sticking out of polyester trousers. None was less than corpulent. From their appearance and the back window, I guessed they were government employees; from their age, it was reasonable to assume that they had fought in the War. The hum was their snoring. They had obviously dined substantially before rolling out their cane mats by the side of the road. One of them seemed to be smacking his chops in his sleep.

Each one was lying flat on his back, facing the sky. A few had arms out, encroaching on their neighbour. They were having nice dreams – of the golden dome over the Imam's body or a cat purring on the ledge. To lie with such abandon in an exposed place – without curling into a ball and turning this way or that – conveyed an impression of profound content.

Amini was standing where we had arranged to meet – at the south end of the Allahvardi Bridge. He had a sportsman's barrel chest and his cheek was scarred by a crescent of beard. He had penetrating eyes but he was taciturn. His face and head didn't seem to fit. They could have come from different kits. (I later learned that his forehead was a collage of metal and bone. He had around sixty bits of shrapnel inside him.)

He held strongly onto my hand, and said: 'Come and meet the lads.' There were about ten of them and they invited me to take off my shoes and sit with them on a rug that had the image of a nine-year-old girl with a perm and blue eyeliner. They ordered two water pipes and tea all round.

The riverside had families eating kebabs that had been cooked on portable grills, and children throwing a ball or playing badminton. The sound of clapping and singing came from the bridge. Some young migrant workers – lads from the province of Lorestan, in western Iran – were having a party and their voices bounced off the water.

Amini and his friends were the depleted shadows of the contented men I had seen sleeping by the side of the road. They were superficially similar: same age, same grey trousers and greying socks, the same semi-beard. But these men had passed over the sinecures and the junkets. (Either that, or they'd been passed over.) They were struggling shopkeepers or small-time *bazaaris*. Their paunches were small. Underneath their jokes, there was a current of shared tragedy – something that I, they told me with bland, distrusting eyes, would be incapable of understanding.

One of them, whose surname was Zahmatkesh, had a vivid red beard, which isn't something you see often in Isfahan. He was sitting opposite me, in a heap that seemed less substantial than his broad torso suggested it should be. Surreptitious investigation revealed that he had no legs; they must have been blown off by a mine. He was the joker of the group. He called me 'Mister' so it rhymed with 'cheetah', and the others sniggered.

'Here, take this,' said Amini, and handed me the snake of one of the water pipes. He put a glass of tea by my side. 'Mr Zahmatkesh has brought a watermelon. You can have some after your water pipe.' Amini was putting sunflower seeds into his mouth, one by one. He would split the shell between his teeth, spit it out, swallow the seed and start again.

We were waiting, he told me, for another friend to turn up; then we'd start talking about the War. A conversation started about degenerate youths. 'I had one at the shop yesterday,' Amini said, 'jabbering into his mobile to his girlfriend. Rich kid, drove a nice metallic blue 206. Offending everyone with his crude talk. He asked for orange juice, and I said, "no". He said, "You don't have orange juice?" I said, "Thanks be to God, I don't have to sell anything unless I want to. At least I've retained that much independence."'

'How do you know it was his girlfriend?' asked someone. 'It might have been his wife.'

'These kids don't have wives,' he replied dryly. 'They have freedom.'

Zahmatkesh addressed me. 'You have freedom in England, don't you?' He was smiling unpleasantly.

I guessed what was going through his mind. Europeans were always lecturing Iran on the importance of freedom, but what did that mean? The freedom to flaunt un-Islamic sexual relations? To parade stolen wealth in front of the mass of the people? To drink alcohol and smoke opium? If that was freedom, Iran had a surfeit. I smiled, trying to convey the impression that I perceived the same irony as Zahmatkesh, and wishing I wasn't British.

The conversation moved on to some recent street protests. They had been held to commemorate a larger, pro-democracy protest a few years before. The protests had been attended by students and other mostly young people who were demanding more political freedom. The participants had been beaten up by the police and hired toughs.

'I was passing,' said one of Amini's friends, 'and saw everything. You should have seen the cops! There were ten for every kid!' It wasn't clear whose side he was on. 'After the demonstration, I saw one cop rip up the seats of a girl's car and beat the shit out of her.'

'Why?' I asked.

'Moral corruption. She had a pink headscarf, halfway up her head, and she was made up like a whore from north Tehran. I think also she may have given him some lip.'

Their smiles were listless and cynical. They were sad, I suppose, that a girl from Isfahan, a conservative provincial town, should dress and make herself up like a kitten from Elahiyeh. They were sad as well that a policeman should beat a teenager with a baton. You could only smile at the absurdity of a state training and employing men to injure its own citizens.

The singing and clapping from the bridge got louder. One of the Lors was dancing sensually in one of the arches, arms above his head, his girlish rump pointing down the river.

'Freedom,' said someone. 'Freedom for them to dance.' Lors did a lot of the menial jobs in Isfahan. They worked hard, for poor wages, and some Isfahanis resented them.

'What's it to us, if he dances?' Mr Zahmatkesh gave a hollow laugh. 'Let him dance.'

Just a few years ago, there would have been no dancing on a bridge over the river in Isfahan. If a man wiggled his bottom in public – not with a woman, just with himself, or male friends – he'd have been arrested. He'd have spent a painful time in a small room at a police station. He'd have got a lecture on Islam's injunctions against depravity. If he went before a judge, he'd have been sentenced to a lashing. A bit further back, say ten years, you'd not have seen much laughing in public. Women had to wear black. Men didn't wear short sleeves. Young people with boyfriends or girlfriends would be severely flogged. No one dared put on Western music in their cars. It felt as though the War was still going on.

From the sneer he was directing at me, it was obvious that Zahmatkesh was preparing another assault. To pre-empt him, I blurted out the first thing that came into my head. 'Is that why Hossein Kharrazi went off to fight the Iraqis? So that Iranian policemen can beat up young women?' Everyone stopped smiling. We listened to the clapping and singing. I felt angry and embarrassed.

After a while, one of the lads got to his feet. He said he should be going. Someone else looked at his watch and said that the man we were waiting for probably wouldn't turn up. One by one, they started to leave. Zahmatkesh bounded like a chimp several metres to the road. By the kerb, there was a contraption made of two motorbikes with a steering wheel on a platform welded between them. He leaped on and roared into the night.

No one offered me a lift back to the hotel. Amini told me to phone him but his tone suggested that he hoped I would not. As I left, I reflected that perhaps they hadn't been waiting for someone else after all; they'd invented him and his non-appearance in order to have an excuse not to talk about the War. Or was I inventing conspiracies of my own?

❖ ❖ ❖

The next day I went to see Hossein Kharrazi's parents. They lived in an older part of the town, with hot, dry streets and high walls casting razor shadows. The corners of the streets had been scraped by fruit carts that were too wide, exposing straw in the daub. I came across a man and asked for directions to Hossein Kharrazi's house. When I found it, I rang the bell and old Mrs Kharrazi's voice crackled over the intercom. When I told her what I'd come for, she called her husband and the door opened.

Mrs Kharrazi told me to come in, rearranging her chador so it wouldn't slide off her head. I took off my shoes and she showed me into a living room that looked out onto a courtyard. There was a big mural on one of the walls; it was a reproduction of a photograph I'd seen before. It showed Hossein in combat fatigues, talking into two microphones that had been taped together.

'Sit down! Sit down!' Hossein's father, a pale old man without a hair on his head, came into the room, grinning. I grinned back, partly because Mr

Kharrazi was wearing pink striped pyjamas. We sat down on a lurid red carpet that covered the floor. I remarked that, apart from the mural, the room contained no sign of Hossein. 'After he was martyred,' Mr Kharrazi said, 'everyone wanted photos of him. Soon, we'd given most of them away. That's why we had this one painted on the wall. No one can cart that off, can they?'

I'd come to see Hossein's parents because the picture I had in my mind of their son was incomplete, puzzling. The outstanding qualities that people talked about – piety, physical courage, modesty and selflessness – seemed to correspond to an archaic Persian ideal of manhood, passed down by the epics. At the same time, the men who had served under him spoke protectively about Hossein, as if he were still vulnerable, more than fifteen years after his death.

During the hour or so I was with them, the Kharrazis gave me a subtly different impression. The Hossein they spoke of needed no protection; he had longed for martyrdom in the service of God and the Revolution. Mr Kharrazi repeated much of what I had learned from official hagiographies: that Saddam had offered a treasure to anyone who could capture Hossein; that Hossein had taken special leave to undertake the pilgrimage to Mecca; that Hossein had been posthumously honoured by top members of the regime.

Mr and Mrs Kharrazi had stopped mourning. 'We're satisfied if God is satisfied,' said Mr Kharrazi. He meant me to infer that God had honoured Hossein by drawing him to his side, and that mortals had no business questioning his will.

Since Hossein's death, the Kharrazis had received top ayatollahs and cabinet ministers in their house. They had, I guessed, enjoyed some of the privileges that are granted to the parents of important martyrs, like discounted trips to Mecca and a place among the bigwigs during the annual commemorations. Mrs Kharrazi sniffed when I asked her whether she mixed with the mothers of other martyrs; she evidently considered herself a cut above them. When her husband mentioned Rahim Safavi, Hossein's old friend who had gone on to succeed Mohsen Rezai as head of the Revolutionary Guard, she said, 'If Hossein had stayed alive, he'd be where Rahim is now.' It made sense to me that the Kharrazis had enlarged

Hossein's public image and put it on their wall. It was the public Hossein, not the private one, who had made them what they were.

While Mrs Kharrazi was preparing tea, her husband passed me a photograph that showed Hossein's head lolling, covered in dust. A trickle of blood was coming out of the side of his mouth. Mr Kharrazi smiled at Hossein's beautiful martyrdom and looked up, expecting me to smile too.

Shortly before I left the Kharrazis, I turned off my tape recorder, and repeated something that I had heard vaguely expressed in Tehran and had wanted to ask Amini about. 'Isn't it true that there was a dispute between Hossein and some important officials, shortly before he died?'

They glanced quickly at each other. Mr Kharrazi said, 'There was no such dispute.' Mrs Kharrazi said, 'Of course, there may have been little problems, but only little ones . . .' Her husband nodded, and said, 'Nothing big, though,' and his smile grew chilly.

❖   ❖   ❖

After Khorramshahr, Saddam Hussein ordered his forces to observe a unilateral ceasefire and Iraq started withdrawing from the remaining areas of Iranian territory in its possession. Encouraged by Iraq, international peacemakers increased their efforts to persuade Iran to negotiate. Some Arab countries, including Saudi Arabia and Kuwait, were reportedly ready to pay Iran reparations on Iraq's behalf.

About the time I first visited Iran, in 1999, people were beginning to question Iran's decision to rebuff these efforts. They drew attention to an interview given by Ahmad Khomeini after his father's death, in which he said that the Imam had opposed the War's prolongation but had been won over by 'those in charge'. In his memoir, Rafsanjani, who was perhaps the Imam's closest adviser, presents things differently. He says that Khomeini wanted the War to continue but that he opposed an invasion. Khomeini apparently disliked the idea of making peace with the invader, but feared the repercussions of an invasion – not least on Iraq's own Shi'as, whom Saddam might single out for reprisals. The two positions were hard to reconcile. Iran had recaptured almost all its territory and Saddam wanted peace. There was little, short of invading, that the Iranians could do to keep the War going.

According to Rafsanjani's memoir, an important meeting was held seventeen days after the liberation of Khorramshahr. 'The military commanders had a discussion with the Imam and demonstrated, with recourse to their professional expertise, the contradiction between continuing the War and a prohibition on entering Iraq. At length, they convinced the Imam that they should enter Iraq from places where there were no Iraqis living, or few living.' But several pages earlier, Rafsanjani implies that the Imam had already been convinced. A mere three days after Khorramshahr, at a meeting of the Supreme Defence Council attended by Mohsen Rezai and Sayyad Shirazi, it had been decided to enter Iraq 'from a very limited number of points, in order to secure our own cities, and to get reparations'.

Again, there are subtle differences between this and the version that Rezai gave when I met him in 2003. He said that neither he nor Shirazi attended the meeting to which Rafsanjani alludes, three days after Khorramshahr. Far from being an initiative driven by military men, he argues that the invasion scheme was equally favoured by 'politicians' on the Supreme Defence Council – by which he means Khamanei and Rafsanjani – who wanted Iran to have cards in hand if and when it came to negotiating. At the end of the second meeting, which Rezai agrees he attended, Khomeini wanted to 'think things over, which some present as evidence for the Imam's opposition. In fact, the Imam had queries and questions, which does not amount to opposition.'

All sides, therefore, disclaim sole responsibility for one of the most far-reaching decisions in recent Middle Eastern history – a decision that would prolong the War a further six years and influence Saddam's decision to invade Kuwait in 1990. By saying that his father opposed the invasion, Ahmad was trying to secure the Imam's posthumous reputation. Rafsanjani pointed at the military commanders; Rezai retorted that there was a consensus between military and civilian officials. In fact, if there was a split, it was between the gung-ho Rafsanjani and Rezai on the one side, and the more cautious Khamanei and Shirazi on the other. Ayatollah Montazeri strongly implied when I met him that Rafsanjani had been the key agent in persuading the Imam to countenance an invasion.

None of this means that the idea was unpopular. Over the eighteen months between Iraq's invasion and the liberation of Khorramshahr, the Iranians had fought against overwhelming odds: against Saddam Hussein and his army; against the will (and arms) of much of the world; internally, against the Mujahedin, in a civil conflict that only started to go the regime's way in 1982. A feeling of invincibility is understandable. Moreover, Iran's sixty thousand dead were sufficient to generate a lobby of bereaved families demanding vengeance. Millions of Iranians still believed that force could be used to export the Revolution. Israel's invasion of southern Lebanon had made this objective urgent; if something wasn't done, the whole region could fall under the shadow of the Zionists.

None of the Revolutionary Guard and Basijis that I have spoken to recalls hearing – let alone voicing – serious criticism of the decision to invade. There was a feeling that Saddam's protestations of peaceful intent couldn't be trusted; if his war machine were left intact, he would be free to attack again. The act of expelling the Iraqis, furthermore, had given a misleading impression of the enemy's weaknesses. Stories circulating after Khorramshahr traduced the Iraqis' virility and religiosity; I've spoken to two participants in the operation who claimed to have heard of or seen the Iraqis sodomizing livestock. Iranians compared their 'compassionate' treatment of Iraqi prisoners with stories of Iraqi atrocities towards Iranian POWs. The cumulative effect of such propaganda was to depict the Iraqis as base and cowardly – no match for Iran's courage and faith.

Having decided not to make peace, the Iranian leadership came to feel that it couldn't discuss terms until it scored another famous victory, of the scale and drama of Khorramshahr. When it eventually came, a tardy two months after Khorramshahr, Operation Ramadan, Iran's first major attempt to invade, was a bloody and costly failure. According to Ayatollah Moussavi Khoiniha, who was public prosecutor at the time, Khomeini suggested negotiations after Ramadan but was dissuaded by the people around him. Even if this is the case, the Imam soon became the exponent of war until the bitter end. Even after September's relatively successful Operation Moharram, there was much less talk of peace.

Some people have a cynical explanation for all this. Khomeini's state, they say, had never been by nature democratic. Now, wartime exigencies were an excuse for absolutism. There was the brutal campaign against suspected Mujahedin sympathizers. In the autumn of 1982, Bazargan's former foreign minister was executed for plotting a coup and Ayatollah Shariatmadari, Khomeini's only rival among the political ayatollahs, placed under house arrest for his alleged involvement. (Montazeri says that he has it on 'reliable' authority that the case against both was 'fabricated'.) The regime went on to crush the Communists who had supported it; their party was banned and ten leaders executed. Freedom of the press and speech had long been curtailed.

The Basijis continued to adhere to their slogan, 'The road to Jerusalem passes through Karbala'. For them, invading Iraq was less a strategic decision than a religious adventure. They dreamed of praying at the tomb of the Imam Hossein, then marching to smash Israel. Without a clear idea of why they were doing it, but convinced that it was God's will, Iran's fearless lions went forward.

❈ ❈ ❈

'Thanks be to God we're back home on Iraqi soil after Khorramshahr. I remember when we went in – I'm a professional, been in the field three years – and wondering how it'll be when we get their oil and a nice long coastline. They told us the people there were just like us. They said the Arabs there hate the Persians; they'd greet us like liberators, offer us their daughters, kill the fatted calf. Did they? They stared like we had our snouts in their shit.

That's a good joke, but the army's the wrong place for jokes that are deficient . . . how shall I put it? . . . in the required bollock rubbing. You never know who's the informant in your unit – I suspect the rat-like Kurd with the suspicious supply of chocolate. (They must have got to him through his sweet tooth.) Remember not to open your mouth or next thing you know the Baathists are taking a bayonet to your ear. A branding: mark you down for the abattoir. Soon enough, you get posted to the front for the Basiji once-over. (Once is the only chance you get.)

A year and a bit since the K calamity. (Makes me tremble to hear the name. Me! Strong as an ox. That's how bad.) I'm lucky to be alive; my past as a divisional swimming champion served me when we tipped into the river. Think of it. Me and some of the lads on a raft, trying to cross with the flaming city behind us. Inflatable raft, holds eight. We're thirteen, fourteen, and they've found their range. Mortar shells, water churning blood and one of the lads loses it next to me, goes off his head and starts hitting and shouting stuff, trying to push us off the raft into the water.

The lads around me, screaming – can't swim most of them – and me, I'm swimming to get away from them, knowing if they grab me I'm dead, they'll take me down. One of the hands from behind, grasping my shoulder. I hit him in the face, feel his nose break. I'm not the superstitious type but that bastard's disturbing me when I sleep, I feel his touch on my shoulder. A week ago my own nails woke me up, scratch-scratching at the old shoulder. Next morning, blood under the nail.

And the Iranians with their 'Allah Akbar!' watching from the other bank, Allah Akbar! as our lads are swept away, swept down the river like scum down a plug. I hate Allah Akbar! Especially Allah Akbar! I used to like saying my prayers. You can't say your prayers without saying Allah Akbar! It doesn't count as prayers. I tried saying my prayers without Allah Akbar! Then I stopped saying them – makes me feel sick. The lack of sleep is getting me down. I don't know about God.

After Khorramshahr I had home leave and prayed for peace. Know my mistake? I thought they were human. We gave them Islam, they don't like that. We gave them Shi'ism. They're idolaters at heart, sun-worshippers, fire-worshippers, whatever. They worship Khomeini and hate the Arabs; they're jealous that we have the sites, Najaf (where Ali is) and Karbala (Hossein), jealous of our culture, that the Prophet was an Arab and his language was Arabic. Jealous, jealous, jealous. Violent and fanatical. Unstable.

Then one day at home Mamma was crying and I got it out of her. Abu-Salah, my elder brother, got put away for subversion. God knows what he did; probably got caught with some tract in his bag. That's what an education does for you. It's a matter of time before they come for the rest of us. Guilt by association. It's the fear that makes you rat on your brother or your dad, the fear that they'll get you because you didn't rat on your kith and kin! Am I surviving or prolonging? No day without fear.

After my leave they assigned me to ordnance. It's lovely being at the back of an army. Tank formations spread across the plain, berms and kilometres of trench and bunker. Lovely officers' trenches: TVs and fridges and no rats or piss-stink. When the attack comes, they tell the tank crews, trap them in the triangles. Then: crossfire, no way out. Like hosing, they say, and it made me think of hosing the mud off the hubs of a staff car. Whoosh.

That was till yesterday. (Ever had the feeling you know something's coming but you don't for the life of you know when? Today or tomorrow or years from now?) They sent me up to the front line. This is it, I'm thinking. That bastard Abu-Salah. That cunt Abu-Salah! I hope they flay him and put him in a bag full of starving cats.

The Iranians must be planning an attack. We know the signs: movement this way and that along the front; heavy traffic from their front lines to the bathhouse – dummies going that way and real men coming back. We've got intelligence their side. (They say the Americans are giving us satellite info too.) And those mullahs can't keep their gobs shut; there's always one mouthing off in Friday prayers about the next 'final offensive'. Pray it's a feint and they attack another sector.

Now it's me at night in a trench in front of the barbed wire and then the minefield with its decapitated palm trees, and behind that the Iranians. Me and my Kalash and some lads I don't know and don't want to know. From Basra, by their accents, I'd say. Why bother passing the time of day? I'm trying to think profoundly, but nothing comes. I call on God: no answer. I just feel empty, and I'm waiting for the emptiness to be filled by terror.

A sourpuss came up to me this morning and asked: you know how I survived Operation Moharram? I gave him a look that said I'm not inter-ested but he told me all the same. He climbed into the pit under the latrine when the Basijis took our forward positions. Climbed in and stayed six hours. Door of the latrine opens and a Basiji comes in. Sourpuss looks up and sees a white Iranian arse, gets it in the eye. A few hours later, our lads counterattack and retake the line. Sourpuss hits himself with a rifle butt and pretends someone knocked him out. Smelling of Basij shit. Holy shit.

I'm not ashamed to say I stored his idea. Not a bad one, if you value life more than dignity. Then, I thought: why is he telling me this? Is he trying to get something out of me, some plan for deserting? They like to kill deserters before they desert – good example to the rest. We at the front line may be expendable, but we have our uses. The longer we hold out, the more Basijis we take with us, the better things will go when it comes to the counter.

The Iranian barrage has started. I have my Kalash and I need a hole. They do that most nights, the barrage I mean, so we don't know when they're coming. Could be tonight. Could be setting out now, the mine-clearers, clearing a path. Basijis following up to the wire. That's when you know for sure, when they blow the wire and it goes leaping like tumbleweed, blowing mines as it goes. I remember once we lit them up with a lucky flare before they had a chance to blow the wire. Shivering, ecstatic Basijis caught in the light. No time for them to lay the explosives – one of them threw himself across the wire so the others could walk across. That mad Basiji didn't make a sound as his chest got cut to ribbons.

And then they come at you, bunched together madly, impossible to stop, just running and shouting running and shouting their blasted refrain: No! Don't say it! Don't think it! Please let it be dawn soon, another day.'

Questions were raised, but they seemed to concern tactics, not strategy and goals. There were bound to be differences of opinion at the front and across society. And yet there was trepidation in the asking, a fear of admitting misgiving into the 'what if' – God forbid, a calling into question? It was in part a reaction to Khorramshahr; the Iranians' fissiparous nature reasserted itself. But there was a simpler worry. It was becoming difficult to suspend the disbelief.

Things had been simpler while the fighting was in Iran. There had been no minefields. There had been the reassuring foreknowledge that the Iraqis would turn green at the first '*Allah Akbar!*' Now, Iran's attacks on Iraq were going nowhere. The invincible ship was becalmed, in a slick. Of what? Doubt? Could it be that Iran's main ally, God, was fed up? Some people had moral pangs; Iran, having confused injury with virtue, had become the aggressor.

You started to find the odd Revolutionary Guard who regretted that the men issuing the orders had not been taught the art of war. There were Basijis who wondered whether zeal and expertise could be combined without compromising the former.* An articulate officer corps might have been able to answer the doubters. But the officer corps, insofar as it existed, was part of the problem. Revolutionary Guard commanders didn't become Revolutionary Guard commanders because they were good soldiers. (Hossein Kharrazi, self-taught, intuitive, had to turn himself into one.) They got there through zeal, loyalty and luck. Most of the expert soldiers were in the regular army or had fled abroad, caught up by the monarchist retreat.

Two images, described to me by the people in them. One is a Revolutionary Guard commander alighting from a staff car near the front and asking a Basiji to take a photograph of him while he loads the Basiji's mortar. The big shot wears pressed trousers and has a trimmed beard and smells nice – this distinguishes him from the Basiji, who has forgotten the meaning of personal hygiene. The commander tries to load the shell in the wrong end of the mortar. The Basiji is disgusted.

The second image was described to me by a saboteur from Shiraz called Abbas. After Khorramshahr, Abbas's ten-man troop got a new leader, a young seminarian. The reason for the seminarian's appointment was unclear. He knew little about war, but he knew it was a good idea to kill the enemies of Islam. (He knew nothing about mines.) His Islam did not smile; he threw cold water over men who tarried in rising to say the dawn prayer. One morning, the seminarian told his men that they'd been ordered to clear a path through a minefield, in preparation for an attack.

The trick is to crouch down and move forward warily, probing the ground ahead of you with the tip of your bayonet. (The bayonet should be as flat to the ground as possible, so it harmlessly strikes the edge of the mine, rather than its deadly face.) If you come across a mine, you must first excavate around it. Then, you remove the charger, which renders the mine harmless. It's as well to hold your breath for much of this process.

---

* Divine intervention even affected tactics. In his memoir, Shirazi recalls an instance when Mohsen Rezai flew to Tehran for advice about a specific engagement and Khomeini advised him to consult the Qoran. Shirazi writes this in admiring, rather than appalled or incredulous, terms.

Calm nerves are essential. From discovery to neutralization, the process takes only a few seconds.

After the lads started clearing the mines, the seminarian stood behind them and muttered about how much time it was taking; the elaborate care and holding of breath seemed unnecessary. (His nerves may have been rattled by the Iraqi guns, which had found their range and were dropping shells around them.) Then: *elhamdullillah*, an idea! Rather than bayonets, the lads would use spades! They would cover more ground quicker. The spade would encompass a wider area than the bayonet.

The seminarian announced his idea. Abbas said: 'With the greatest respect, you must be joking. You can't find a mine with a spade. You can't control the end of the spade the way you can control the end of a bayonet. A spade is too heavy and blunt.' The mullah demurred; using a spade, he said, would be safer as well as more effective; with a spade, you were further from the mine. Abbas shouted: 'I'm not going to use a spade!'

There were some young lads among the mine-clearers, boys of fifteen and sixteen, and they decided to obey the seminarian. The seminarian refused Abbas's entreaties to revoke the order. Spades were distributed and the lads got to work, with the seminarian leading the way. Shortly after, there was an explosion. Everyone looked up. One of the lads had blood spurting out of his eyes and stumps instead of arms.

The golden rule of mine clearing is: no arguments. If there's so much as a minor verbal contretemps, the troop leader must order his men out of the minefield; he lets them back after they've cooled off. (You can't clear mines if you're perturbed or short of breath.) Abbas begged the seminarian to call a halt. The seminarian threatened to shoot anyone who rose from his job. The lads carried on working and they carried on getting blown up. One of them threw down his spade and went crazy walkabout, lurching erratically. He was pursued by another. They lost a leg apiece.

Abbas's experience was being replicated; it had to be, in a gigantic army that prided itself on its ignorance of military affairs. Some people treated these experiences as they might a letter they suspected of bearing bad tidings – slipping it into a pocket for opening later. Some took tangential exits. They allowed unrelated subjects to become thoughts, and these thoughts acted as a diversion. Family, money worries, mundane stuff that had nothing to do with the War. Some of them decided they didn't want to die.

A few, often the educated and self-confident ones, tried to devise ways to make things better. There were books available in Persian; one book, written by a Vietnamese general, observed that America owed its defeat in the Vietnam War to the Vietnamese irregulars' transformation into a proper army. Some read about the Western Front, another slow-motion war waged for a few metres of blasted mud. Previously, they'd assumed that everything about their war differed from everything about every other war that had ever been fought – except, of course, the early wars of Islam.

Most of the lads, hundreds of thousands of them, preferred to deny that anything was wrong. They saw what they wanted and carried on declaring – to themselves, to their comrades – the inevitability of victory.

Some of the doubters decided, individually and among friends, that restructuring was needed. Even the Revolutionary Guard needed a hierarchy – to impose discipline and coordinate action. The problem of training should be addressed, too. You had Basijis who were incapable of thinking on the battlefield. They didn't know – perhaps they didn't care – that measures existed that would increase their chances of staying alive. Somewhere, through the pride and obfuscation, there ran a line – between what was moral and Islamic, and what worked. That line ran from the front all the way to Tehran, through Khomeini's office, the President's office and the Foreign Ministry. It affected everything.

Take the Arabs. The Basijis despised the Saudis and Kuwaitis who provided the Iraqis with their excellent tinned food. They regarded the Gulf Cooperation Council, which the six southern Persian Gulf states had set up in 1981, as a blatant anti-Iran alliance. Some of the Arab states were providing generous loans to Iraq; all of them lent a diplomatic hand. The Saudis and Kuwaitis – dangling oil at pre-war prices – were instrumental in persuading France to deliver to the Iraqis dozens of Mirage fighter-bombers – warplanes of the kind that Iran could only dream of buying.

The Basijis loathed the French, not only because of the Mirages but also for loaning to Iraq the Super-Etendards whose Exocet missiles threatened the export of Iranian oil. (The French added insult to injury by embargoing three miserable missile boats that had been destined for Iran.) The fallacy of French neutrality was conceded by none other than President Mitterrand, when he said, in 1982: 'We do not want Iraq to lose the war ... the age-old balance between the Arab and the Persian worlds must absolutely be maintained.'*

The Soviets climbed the Basiji hate list. At the beginning of the war, they had penalized Saddam's irredentism by imposing an arms embargo on him. Then, in the wake of Khorramshahr, when it looked as though Iran might actually win the War, they had a change of heart. (This coincided with Khomeini's repression of the Iranian Communists, and the advocacy by some officials of intervention in favour of the Afghan resistance movement against the Soviets.) In the second half of 1982, the Soviets lifted their embargo. They gave Iraq access to tanks, fighters and missiles. By the end of the year, at least one thousand Soviet military advisers were in Iraq.

In 1983, the Basijis' loathing returned to its original object. That was the year that the Reagan administration allowed increased sales of US equipment to Iraq, including sixty helicopters for 'agricultural use'; it also intensified its diplomatic efforts – christening them 'Operation Staunch' – to convince black market dealers, and countries making American weapons under licence, not to sell under the counter to Iran. The following year, President Reagan put Iran on a list of countries that 'provide support for acts of international terrorism' – he removed Iraq from the same list – and beefed up the sanctions that Carter had imposed on Iran.

The Iranians had more oil revenue than Iraq, and their lads were queuing up to die, but they couldn't buy decent weapons. The West and Warsaw

---

* The French love affair with Saddam Hussein is related in Kenneth Timmerman's *The Death Lobby*, an account of the West's arming of Iraq before the 1991 Gulf War. Timmerman begins his book by mirthfully describing Saddam's visit to France in 1975, when his host was Jacques Chirac, the then prime minister. Timmerman relates Saddam's delighted reaction to a display of Provençale bullfighting – he rewarded three of the young matadors with cheques worth $200,000. At the end of Saddam's visit – during which the Iraqis placed an order for a French nuclear reactor that was capable of breeding bomb-grade plutonium – Chirac effulged that French policy was dictated 'not merely by interest, but also by the heart'. He presented Saddam with two harlequin costumes for his daughters that the unfortunate girls were obliged to don when Chirac visited Baghdad in 1976. 'Over the next fifteen years,' Timmerman writes, Saddam 'would spend $20 billion on French arms.'

Pact countries were off limits – officially, at least. Iran was forced to rely on middlemen who often turned out to be fraudsters. Dealers charged astronomical prices for inferior or obsolete equipment. Iranian pilots soared in British Phantoms purchased before 1979, but their missile chambers were empty. In time, the paucity of ordnance stopped mattering; the planes themselves were grounded for lack of spare parts. By mid-1983, Iran's operational air strength had fallen to below one hundred warplanes, from four hundred at the time of the Revolution; Iraq had more than three times as many.

For some of the lads at the front, it was natural that the world should hate them. It was a price worth paying for their beautiful endeavour. The Imam had promised to export the Revolution to the whole world. It wasn't diplomacy of the cringing, cavilling, type; it was militant truth. It had to spread – because God and the Imam demanded it – either violently or diplomatically. Iran's embassies abroad were propagandizing, funding, assassinating, stirring up. Feathers would be ruffled along the way.

To others, it was self-defeating. The Bahrainis, whose Sunni monarchy was threatened by its majority Shi'a population, accused Iran of masterminding a coup attempt. The Saudis were furious when Iranian pilgrims turned the annual pilgrimage to Mecca into a grandstand for anti-American (and anti-Saudi) sloganeering. The Soviets were put out when Iran launched a campaign to help 'oppressed Muslims in the USSR'. In Lebanon, where the Revolutionary Guard had established itself soon after the Israeli invasion, two Iran-sponsored groups, Islamic Jihad and Hezbollah, started kidnapping American civilians, one at a time, and blowing up American soldiers, hundreds at a time.

At the front, the lads knew that, around the world, people were acting on their responsibility to spread the Revolution, and that this was right. But it was obvious that the world had united in a strategic alliance to prevent Iran from winning. Given the injustice, some wondered about the morality of expediency, about compromise – short-term concession for long-term strategic gain. If the end of innocence is the beginning of politics, this was it.

While I was researching this book, I happened to read *Homage to Catalonia*, George Orwell's account of the seven months he spent, in 1936 and 1937, fighting on the side of the Spanish Republican government against Franco's Nationalist rebels. The book is Orwell's dispassionate narration of events in which he had played a passionate part. Reading it helped me in the writing of this book, for *Homage to Catalonia* isn't just about a civil war in Spain; it's about the betrayal of revolutions.

There's a congruity between Spain in 1937 and Iran in the early 1980s. Orwell's account of his induction into the militia of the *Partido Obrero de Unificación Marxista*, an ill-disciplined rabble propelled in the same direction only by political ideals – definitely not by deference to the higher ranks – could have, give or take a hammer and sickle and a slug of wine, been written by a Basij inductee. Orwell experienced the same boons as the Iranian revolutionary: 'Many of the normal motives of civilized life – snobbishness, money-grabbing, fear of the boss etc – had simply ceased to exist.' From the equality of pay and conditions between officers and men to the dearth of decent equipment, the Spanish militiaman and the Basiji knew the same pride and privations. Even the generosity of the working-class Catalan – 'if you ask him for a cigarette he will force the whole packet on you' – could have been culled from recollections of 1980 Khuzistan, rather than the Aragon Front four decades before.

The limits of these congruities are as revealing as their extent. While 'meaningless bullets' sing over a stagnant front, Orwell frets at the inaction. 'Against machine guns and without artillery,' he writes, 'there are only three things you can do: dig yourself in at a safe distance – four hundred yards, say – advance across the open and be massacred, or make small-scale night-attacks that will not alter the general situation.' In Orwell's experience, the first was the norm and the third the exception. He adumbrates the second option casually and for the record. Nowhere, despite his great boredom, does he suggest that the militias should advance across the open and be massacred.

Orwell's bland words impressed on me the novelty of Basiji tactics. Like the Spanish militiamen, the Basijis faced machine guns and had little artillery. By encouraging the Basijis to advance across open ground and get massacred, Iran's military leaders violated an accepted article of war, the tactical concern for the lives of one's own men, on a scale unheard of since the Western Front. Think of the totemic status conferred on earlier violations, because of their very rarity: the Charge of the Light Brigade, for instance (and that happened by mistake). The Khuzistan Front was all the more shocking because that strategic concern was, officially and with fanfare, turned on its head. According to the tactics favoured by the Revolutionary Guard, a calculated disregard for the lives of the men became the norm. It was the main subject of glorification in the official panegyrics. The more horrific the circumstances of a man's death, the more futile his expiry, the more acres of apartment block wall would be dedicated to his memorial. If he died a prepubescent, so much the better; Iran's best-publicized martyr was a beardless boy who threw himself before an oncoming tank. And the official view, while simplifying grossly the complicated feelings of the bereaved, did accurately reflect one aspect of these feelings. Amid the grief, there was joy. In France and Britain in 1916 there was just grief.

Orwell inadvertently sheds light on the main difference between the Spanish militiaman and his Iranian equivalent of the future. 'It struck me,' he writes, 'that the people in this part of Spain must be genuinely without religious feeling – religious feeling, I mean, in the orthodox sense. It is curious that all the time I was in Spain I never saw a person cross himself; yet you would think that such a movement would become instinctive, revolution or no revolution. Obviously the Spanish catholic church will come back (as the saying goes, night and the Jesuits will always return), but there is no doubting that at the outbreak of the revolution it collapsed . . .'

No matter how Socialism affected the way Orwell's brothers in arms lived as soldiers, it didn't affect the way they died. Going into battle, they became little different from the Fascists facing them. Fear; anger; blood lust; the instinct for self-preservation; such impulses reduce to an essence in battle, making one soldier like the next. Whatever social utopia they sought, Orwell and his comrades were fighting to bring it about here, now. Although many were prepared to die trying, they preferred the idea of both changing the world and being around to enjoy the fruits. (When shot through the neck, he experienced, with Orwellian phlegm, a 'violent resentment at having to leave this world which, when all is said and done, suits me so well'.)

Not so the Basijis. The confluence of religion and politics in their thin frames was a coming together of personal and universal interests. They, too, were fighting for earthly rewards, but that was not the sum of it. They were neither atheists nor disenchanted Roman Catholics, but dedicated eschatologists. No matter how fine the paradise they created in this world, it would be banal compared to the pleasures awaiting them in the next. Whenever a Basiji threw himself onto a rocket-propelled grenade that had landed in a crowded trench – and such acts of heroism were commonplace – he was doing his comrades the ultimate service. Yet he was doing himself a service, too. He died smiling because he had convinced himself that going in a certain way, with certain words on his lips, would be in his own interest.

Homage to Catalonia encouraged me to enquire about another aspect of war that interested Orwell. It's natural enough for a conventional soldier, fighting a conventional war, to be perturbed when he finds that the civilians in his hometown are indifferent to his sacrifices. Life has continued; the privations of soldiers may seem distant. But how much greater must be the disappointment of the revolutionary soldier, fighting less for the title deeds of earth and cities than for the raising of a new society. Imagine his pain when he comes home and discovers that the people in whose name he is suffering no longer share his beliefs – or, at least, no longer act on them. The soldier asks himself: why am I fighting? Why are my friends dying?

Consider Orwell's first impression of Barcelona, in December 1936:

The revolution was still in full swing . . . waiters and shop-walkers looked you in the eye and treated you as an equal. Servile and even ceremonial forms of speech had temporarily disappeared. No one said '*Señor*' or '*Don*' or even '*Usted*'; everyone called everyone else '*Comrade*' and '*Thou*', and said '*Salud!*' instead of '*Buenos Dias*'. Almost my first experience was receiving a lecture from a hotel manager for trying to tip a liftboy. There were no private motorcars, they had all been commandeered, and all the trams and taxis and much of the other transport were painted red and black. The revolutionary posters were everywhere, flaming from the walls in clean reds and blues that made the few remaining advertisements look like daubs of mud. Down the Ramblas, the wide central artery of the town where crowds of people streamed constantly to and fro, the loudspeakers were bellowing revolutionary songs all day and far into the night. And it was the aspect of the crowds that was the queerest thing of all. In outward appearance it was a town in which the wealthy classes had practically ceased to exist. Except for a small number of women and foreigners, there were no 'well-dressed' people at all . . .

Contrast this with his return to the same city after just three months at the front.

Things were returning to normal. The smart restaurants and hotels were full of rich people wolfing expensive meals, while for the working-class population food-prices had jumped enormously without any corresponding rise in wages. Apart from the expensiveness of everything, there were recurrent shortages of this and that, which, of course, always hit the poor rather than the rich . . . Previously in Barcelona I had been struck by the absence of beggars; now there were quantities of them. Outside the delicatessen shops at the top of the Ramblas gangs of barefooted children were always waiting to swarm round anyone who came out and clamour for scraps of food. The 'revolutionary' forms of speech were dropping out of use. Strangers seldom addressed you as *tu* and *camarada* nowadays; it was usually *señor* and *usted*. *Buenos Dias* was beginning to replace *salud* . . . a small but significant instance of the way in which everything was now orientated in favour of the wealthier classes could be seen in the tobacco shortage . . . theoretically, the government would not allow tobacco to be purchased abroad, because this meant reducing the gold reserves, which had got to be kept for arms and other necessities. Actually there was a steady supply of smuggled foreign cigarettes of the more expensive kinds, Lucky Strikes and so forth, which gave a grand opportunity for profiteering.

Substitute Tehran, circa 1981, for Barcelona. Despite the widespread observance of Islamic rules that prescribe for women a subservient status, the men and women of the Revolution have achieved a kind of moral equality. (They manned the barricades together, they face Saddam and the Mujahedin together.) The revolutionaries have abandoned the old forms of address; 'sir' and 'madam' have been replaced by 'brother' and 'sister'. The oleaginous greetings of pre-1979 Tehran have given way to rigorous proletarian salutations: 'Don't be tired!' 'God's power (be with you)!' In the streets, black has become monolithically fashionable – in honour of the War dead, in tenacious mourning for the twelve Imams. Women have dispensed with the perfumes that once lay surprisingly over the sweat and fumes of the city. There is less eyeliner and lip gloss and the uptown salons have closed. The women's faces are like blank pages. A virtuous frown is the expression of choice – it speaks of piety, severity and benevolence.

When two male supporters of the Revolution meet in the street, they kiss on the upper arm in echo of the Imam's declared desire to kiss the upper arm of every Basiji. (The upper arm is where the fighting man's strength lies; to kiss it is a sign of admiration of that strength.) The men have *kaffiehs* around their necks. You may have trouble identifying their social background. Humble shopkeeper or wealthy *bazaari*? It's impossible to tell, for the *bazaaris* who helped bring Khomeini to power eschew the ostentation of the Shah's upper class. Their daughters still marry the sons of other rich *bazaaris*, but the wedding ceremonies are piously modest, almost sombre affairs. There are, naturally, no ties; the old loud silken marking has all but disappeared. The Cadillac is locked in the garage, the family Paykan being more appropriate to the times.

Having met at university, revolutionary newlyweds settle in down-at-heel neighbourhoods. Inspired by the frugality of the early Imams, they live uncomplainingly in garrets that roast in summer and freeze in winter. In such neighbourhoods, it's advisable to keep noise and laughter to a minimum. There are daily memorial services for local lads. The Imam has banned music, except for religious laments and marches on the radio. (Strangely, you also hear snatches of Beethoven's Fifth, whose energizing qualities outweigh, the authorities have deemed, its regrettably distant origins.)

'Every war', writes Orwell, 'suffers a kind of progressive degradation with every month that it continues.' That degradation took longer in Tehran than it had in Barcelona. (In Iran, the government stood for a while against it; the Spanish government, being directed by people who wanted to co-opt the bourgeoisie, encouraged it.) Sartorially, for example, Iran didn't experience anything like the reversal to pre-revolutionary ostentation that Orwell witnessed in Barcelona. But keeping up appearances didn't mean that the core of the enterprise was still healthy. That, in fact, had begun rotting soon after Khorramshahr, and its effects were obvious to the lads returning from the front – to people like Haji Pakdel.

Haji belongs to Mohsen Makhmalbof, one of Iran's best post-revolutionary film makers. In Makhmalbof's great film, *The Wedding of the Blessed* (1988), which is set during the second half of the War, Haji is the anger and conscience of his creator – and of all Iranians who chafe at the subversion of revolutionary ideals, particularly at the weakening resolve to help the downtrodden. Haji's gaunt, unsmiling handsomeness makes him a kind of idealized Old Testament spokesman for the unattractive Makhmalbof who comes across as a thick-lipped fanatic with accusing spectacles.*

The Wedding of the Blessed starts with Haji's discharge from a clinic for soldiers with post-traumatic stress disorder. At the beginning, there's a tremendous battle scene; bedpans are used as helmets and crutches as RPG launchers while the camera glides in hellish circles around the white ward. But even allowing for Haji's precarious mental state, there's a good reason to suppose that he'll recover: his fiancée, the lovely Mehri. Mehri shares Haji's ideology, though she is tougher than he is. Her flaw is her father, the venal *bazaari* Shaban.

Shaban is the Imam's mistake. It was the bazaar that persuaded Khomeini to relax measures to prevent infringements of government price controls. Prices soared, only dipping when the Imam made a personal appeal to bazaar leaders. Pro-*bazaaris* in the Council of Guardians shot down a bill to nationalize foreign trade; this heralded the bazaar's involvement in the lucrative importation of consumer goods. The Imam even invited the Shah's exiled business elite to return: 'No one,' he declared, 'should be afraid to come back and engage in business . . . as long as there is Islam, there will

---

* After the revolution, Makhmalbof suggested that a special tribunal be set up to 'try' the work of filmmakers under the Shah – the suggestion was not acted upon.

be free enterprise too.' (In *The Wedding of the Blessed*, Haji makes caustic allusion to this magnanimity, saying, 'the devils are returning'.)

None of this amounted to Reaganomics; the government kept control over the biggest sectors of the economy and rationing ensured that few people went hungry. Worse than the new fortunes was the smell of betrayal. When the commerce minister, a *bazaari* himself, had rice de-rationed, the price trebled. (The government swiftly overturned the decree but the damage had been done.) Many *bazaaris* were notorious hoarders – holding onto butter, for instance, until prices rose, and then releasing stocks. According to the government, there were empty properties available for Tehran's homeless hundreds of thousands, but property-owners preferred empty houses to lower rents. I have heard of the wives of Revolutionary Guards being evicted because they couldn't afford a rent rise.

Shaban's status within this new Iran is evoked during the nocturnal drive home from the clinic that he, his family and the discharged Haji undertake. Sitting in the back of Shaban's Mercedes, Haji views Tehran through the triple-pronged motif on the bonnet – a kind of roving rifle sight that alights on fading slogans on walls, before moving on. 'We will drag all capitalists to the court of justice,' promises one bitter old graffito, caught in Shaban's headlights. A second announces: 'The country belongs to the shanty dwellers.' No, it belongs to Shaban.

'In the first months of the revolution,' Orwell writes in *Homage to Catalonia*, 'there must have been many thousands of people who deliberately put on overalls and shouted revolutionary slogans as a way of saving their skins.' The description applies to Shaban. His five-day stubble and shapeless black clothes are blandly revolutionary, but the man's a sham. He lives in a superb mansion of Qajar dimensions – with his avaricious wife and a mentally retarded son who is forever blowing soap bubbles. Thanks to this son, we get to see a photograph of Shaban before the Revolution, enjoying a glass of arrack with a pal. He is as much a materialist as he is a sinner. (In Haji's mind, the two are interlinked.) When Shaban reluctantly acquiesces to Mehri's marriage to Haji, his pal urges: 'Your daughter is a done deal. Let the buyer worry about the profit and loss.'

No ideology, however catholic, could incorporate Shaban and Haji. If anything, the time Haji spent at the front – an experience that persuaded him to reject materialist values in favour of spiritual ones – has imposed on him an even more rigid understanding of his mission. Despite the apparent constraints of the Iranian Revolution, confined as it is to the renascence of Shi'a Islam, Haji is in fact a globalist. He suffers the internationalist's frustration, impotent in the face of conquering world events.

Trying to arrange marriage papers in the notary's office, he's forced to listen in to the bartering of speculators over the confiscated mansions of the Shah's departed cronies. For a while, he teams up with his old business partner in their portrait studio, but his disillusionment and periodic fits don't allow him to sustain any activity.* Returning to documentary photography, his lens – and that of Mehri, who is also a photographer with a social conscience – finds chattels on the streets of the new Iran, not human beings. There are vagrants blinking in the torchlight; a prematurely aged woman raising her hands in exhausted surrender; kids smashing open the public telephones so they can steal coins, one of them misusing that revolutionary totem, the *nanchiko*, for nefarious ends. It is a grotesque distortion; can Haji alone see it? The newspaper prints his ironic image of a sunflower, but not the beggar woman's vacant mirth.

In the end, it's the universal part of Haji's vision, his depression in the face of images of starving African children, which Mehri cannot share. Ignorant of the collectivist atmosphere of the front, she retains an idea of herself and her fiancé as individuals. She is willing to bend, while Haji cannot; he will snap. He goes berserk at his own wedding, the wedding of the blessed, castigating Shaban's friends with raving effrontery for eating food that, in his view, belongs to the poor. Reconfined after the ceremony,

---

* The brief scene of Haji in his studio is one of the most memorable in any Iranian film. A young couple come in to have their portrait taken. We see them upside down, through the prism of the studio camera. We are first surprised, then shocked, to see the girl bring her hands up to her headscarf and start taking it off. It's hard to exaggerate the suspense generated by this gesture, in a country where the enforced covering of women's heads is the regime's most apparent symbol – the symbol, too, of its intransigence. Questions clamour the viewer's mind: isn't that illegal? How did the censors permit such a violation of Islamic law? What is Makhmalbof playing at? And then, as in a striptease, the scarf comes off, and the girl is revealed to be quite bald. Of course! It's not the female head that Islam finds wanton, but the hair. Makhmalbof chuckles.

he escapes from the clinic and telephones Mehri to say goodbye. It goes without saying that Haji is returning to the front, the only place where he can forget.

❁　❁　❁

In 1986, the Iran–Contra scandal broke. In America, there were, in fact, two affairs, divided by a dash and linked by morally ambiguous Cold Warriors like Colonel Oliver North, a pious self-publicist who had been seconded to Reagan's National Security Council. Abetted by North, two national security advisers, Robert McFarlane (1983–5) and Admiral John Poindexter (his successor) subverted the official policy of not selling arms to Iran, a country that America had named as a sponsor of terrorism. The arms were sold, furthermore, in order to win the freedom of American hostages being held in Lebanon by Iran-backed Hezbollah. Didn't that amount to doing deals with terrorists?

Over fifteen months of secret contacts between the Iranians on one hand and the Israelis and the Americans on the other, at least two thousand anti-tank missiles and some two hundred and thirty-five ground-to-air missiles were transferred to Iran. (In return, Iran pressured Hezbollah into releasing three of the Americans they had taken hostage.) In May 1986, an American delegation led by McFarlane took the considerable risk of going (incognito) to Tehran in the vain hope of speeding up hostage releases. The question was: had the president known? Reagan, having first said he'd authorized the shipments, then said he hadn't; eventually, he said he couldn't remember whether he had or not.

Near the end of 1986, a memo was discovered which referred to the diversion of revenues from the Iran sales to the Contra rebels who were then fighting to topple Nicaragua's Soviet-backed government. A second question arose: had the president authorized an apparent violation of a recent law that prohibited the allocation of appropriated money to the Contras? The answer turned out to be something like 'not in so many words'. Poindexter, as he made clear in the public hearings that followed the revelations, had kept Reagan in the dark about some aspects of the Contra file, admitting his boss to the rickety moral redoubt of 'deniability'.

(Poindexter and the other main fall guy, North, certainly had the impression that they were doing things the president would have approved of.)

For all the intimidating details and complexities of what turned into a slow-motion slugging match between Congress and the executive, the Iran–Contra affairs were about simple things: means and ends, and the deception of the public. (And God; North, a believer in the sanctity of the anti-Communist cause, brought him a lot into things, and everyone knew that Reagan was receiving 'guidance'.) In the end, legal technicalities and waning public appetite for heads on stakes meant that a grand trial with grand charges was reduced to a process of petty felons, with proportionately risible sentences.

In Iran, there were executions, and they were ostensibly to do with the purchase of arms, under the table and at insultingly inflated prices, from Iran's worst enemy, Israel. In fact, as soon as it became clear that Khomeini had endorsed the arms purchases, all need to account for the deception evaporated. You don't refer to someone as 'the Imam' and reserve the right to question him. On the contrary; if you see something that confuses or upsets you, you put it down to your own ignorance. That's what most Iranians did.

There was a flaw, however – an Iranian who stood grinding his axe right in the belly of the regime: Mehdi Hashemi. It was Hashemi who put an account of the McFarlane trip at the disposal of a Lebanese magazine, which announced it to the world. Before long, American ideas of right and wrong had distracted the Western media from the dimly discernible tensions of the Iranian Revolution – a miasma of contradiction and ambiguity at the best of times. And yet, what happened in Iran was to be much more defining of that country's future than Iran–Contra was to be of America's. In Iran, it wasn't a dusty quadrennial that cracked but an eternal Revolution.

The previous year, the Assembly of Experts – a kind of Iranian College of Cardinals – had named Hashemi's protector and sponsor, Ayatollah Montazeri, as Khomeini's eventual successor. Until then, the Revolution had been Khomeini and Khomeini had been the Revolution; no provision had been made, either in the constitution or the popular imagination, for his passing. It went without saying that no one could succeed him as father of the nation. The most that could be expected of his successor was competent leadership (albeit for life). Montazeri had been the obvious choice from a restricted field; of the three Iranian theologians with the required erudition, he alone had the necessary revolutionary refulgence. Although his candidacy was supported by Rafsanjani and President Khamanei, and endorsed privately by Khomeini, the Assembly deliberated for more than two years before finally anointing him. Then, his position became analogous to that of an American vice-president. He was one supposition from exceptional power. In the meantime, he was no one.

Even after he'd been named as Khomeini's successor, and his genial, peasant-like face started appearing on TV, there were doubts. Montazeri's reputation for blunt talking bordering on recalcitrance, and his status as a spectator of tumult and events, made him the natural figurehead for critics of the War. As his own opposition to the War hardened in 1984 and 1985, he received visits from dozens of Revolutionary Guards, who complained that victory and lives were being thrown away. Their litany of rumbled operation plans, bad command decisions and defeats dressed as triumphs found a willing ear. Sometimes, Montazeri passed on these complaints, in the form of letters, to the Imam's Office. Occasionally, he lodged misgivings in person. Rarely did he feel as though people were listening.

In different circumstances, the regime might have made something of Montazeri. He could have been a lightning conductor, neutralizing criticism before it did proper damage. But that would have needed a good working relationship between himself, on one hand, and Rafsanjani and Ahmad Khomeini, on the other. The clergy of Qom is a formal society that pays elaborate courtesies to its most senior minds; Montazeri might have felt slighted by the indifference shown him by Khomeini's two chief courtiers – clerics of inferior standing. More generally, he resented the influence that Rafsanjani and Ahmad enjoyed over Khomeini. They controlled the flow of information and supplicants to the Imam's Office, and gave him, in Montazeri's view, a sanitized, even dishonestly sanguine, account of events. According to Montazeri, they did their best to prevent commanders from conveying their unhappiness to the Imam. Rafsanjani, he implies, didn't want the Imam to know how badly the War was going.

Rafsanjani had taken control of strategy and sometimes tactics. 'Mohsen Rezai,' Montazeri says, 'would telephone [Rafsanjani] from the front line and [Rafsanjani] would order him to do such and such, or go to such and such a place. Now, I don't know whether the Imam gave [Rafsanjani] the authority to do this, or whether he just took it upon himself.' Montazeri says he was in the position of having to defend the very 'gentlemen' – an ironic reference to Rafsanjani and Ahmad – whom he held responsible for Iran's battlefield failures. 'I would justify the errors and mistakes and delusions of the gentlemen and top commanders. At the same time . . . I would explain to the gentlemen the bitter reality and point out their delusions, but they took no notice.'

Montazeri and his entourage suggest another reason for the personal antagonism that poisoned the atmosphere between the Imam's secretariat, in Tehran, and his successor's, in Qom. According to Montazeri and his entourage, Rafsanjani and Ahmad Khomeini doubted their continuing ability to direct the nation after the succession. They feared that they would be replaced, as advisers and confidants of the most powerful man in the country, by Montazeri's secretariat. They wanted, according to this account, to supplant or eliminate Montazeri's trusted advisers. There were obstacles to their purpose, however, and the most formidable was Mehdi Hashemi.

Everyone agrees that Hashemi was gifted – the kind of man, according to an admirer, 'who could issue operational orders destined for Islamic militants in the Philippines, reply by hand in elegant Arabic to a letter from Yasser Arafat, listen attentively to the BBC Persian service and read the newspapers at the same time'. He was the quintessential Iranian revolutionary – a blur of atavism and industry, and capable, naturally, of slewing towards violence.

Shortly after the Revolution, Hashemi had been part of an armed force that tried, while Bazargan was prime minister, to board a civilian Iran Air flight bound for Lebanon – which boisterous act embroiled the force in a sensational standoff with police at Tehran's Mehrabad Airport. Saeed Montazeri, the ayatollah's second son, casually attributes this behaviour to the 'atmosphere of extremism that was prevalent in those early days'. In 1981, this atmosphere, combined with his talents, propelled Hashemi to the leadership of the Office for World Islamic Liberation Movements. It was a cumbersome name for an export promotion authority with an interest in Islamic revolution. It had no greater supporter than Ayatollah Montazeri.

When I met him in 2003, the delightful and frail Ayatollah Montazeri had been packaged as a moderate, the smiling (and pluralist) face of revolutionary Shi'ism. I remember a statement that he released in 2002, in which he implicitly argued for a two-state solution to the Middle East crisis – as long as the putative Palestinian state had Jerusalem for a capital and the Occupied Territories inside its borders. Such an arrangement, Montazeri said, would suit not only the Palestinians; 'it will be to Israel's advantage, too'. The statement was a repudiation of one of the Islamic Republic's shibboleths – its refusal to accept Israel's right to exist. It didn't mean that Montazeri liked Israel's existence. If, he was saying, in return for accepting Israel, Islam could get back Jerusalem and the Occupied Territories, it might be worth doing a deal.

Such *realpolitik* is a far cry from the Montazeri of the early 1980s. Remember his son's comment about the 'atmosphere of extremism'. Remember, too, the stark, primary colours – the innocence of that extremism. Clerics whose perception of abroad was shaped by prejudice and half-truth had set up a foreign policy that had nothing to do with diplomacy and everything to do with the divine necessity of imposing one's beliefs on others. Noting the flaccid Islamism of Iran's diplomats, Montazeri set up a university for the envoys of the future. Responding to Arafat's call, he tried to set up a faculty for Palestinian kids that would obviate the need for their going to Communist countries for higher education. In 2003, he told me that his understanding of exporting revolution 'didn't amount to Iran's military intervention or terrorist actions in other countries; on the contrary, it was about publicizing . . . the genuine culture of Islam in the face of all sorts of dictatorship and repression'. But that, it seems, was not the whole story.

Take Hashemi's Office for World Islamic Liberation Movements. When I spoke to him, Saeed Montazeri dwelt suavely on its 'educational and propaganda' functions, but the Office was part of the Revolutionary Guard and its function was primarily military. Islamic militancy was a growth sector in the 1980s. Palestine and Lebanon, Afghanistan and South-East Asia: Hashemi's boys were active in all of them. They are said to have had a camp in Syria, where would-be ideologues from several countries got training and indoctrination. They had good ties with Libya's Colonel Gaddafi. In his memoir, Montazeri refers cryptically to collaborative 'activities abroad' with the Intelligence Ministry. When I asked Montazeri about these activities, he spoke blandly of 'cultural and academic work with the young people of oppressed nations', but elements of the ministry, by the time he wrote his memoir, had become notorious for its involvement in the assassination of Iranian dissidents overseas.

In theory, Hashemi worked under the Revolutionary Guard. In reality, the Office had its own finances, contacts and mission. Some operating costs came from Montazeri, who received untold sums annually from his followers, in the form of tithes and Islamic taxes (tax-free, of course). These were topped up with donations made directly to the Office's account at one of the big banks. Even before Montazeri was appointed Khomeini's successor, his protection, and the Office's financial independence, allowed Hashemi to manage his relations with Hezbollah and the Afghan Mujahedin without reference to higher authority. In some Revolutionary Guard units, Hashemi's popularity exceeded Rezai's, and his name was invoked when some units in Tehran started an agitation to have Rezai dismissed. (These units, according to Saeed Montazeri, were expedited to the front and put in the van of a reckless attack that decimated them.)

The Foreign Ministry disliked Hashemi for trying to overturn a tentative détente that it had initiated with Arab countries. In 1984, parliament, efficiently marshalled by Rafsanjani, the speaker, removed responsibility for the Office from the Revolutionary Guard and handed it to the Foreign Ministry and Intelligence Ministry. Shortly after, the Office was wound down; Hashemi himself was sidelined, in the words of one of his enemies, 'with the greatest difficulty'. Even then, Hashemi and his lads carried on trying to export the Revolution and conflict with the Foreign Ministry went on.

Nowadays, the Montazeri camp rejects their opponents' claims that Khomeini had distrusted Hashemi all along; they finger Rafsanjani as the prime mover to have him sidelined. Rafsanjani's supporters, on the other hand, maintain that the Imam himself dispatched senior clerics to persuade Montazeri to distance himself from Hashemi. They cite an account, allegedly related by the head of Intelligence inside the Revolutionary Guard, of a much earlier warning from Khomeini to senior commanders, to 'watch out for Mehdi Hashemi'. Montazeri's supporters accuse their opponents of falsifying evidence; the Imam, they say, never issued such a warning.*

Sometime towards the end of June 1986, one of Montazeri's associates brought him copies of two letters that had purportedly been written by Manuchehr Ghorbanifar, an Iranian arms dealer, to Mohsen Kangarlu, who was in charge of procurement for the Revolutionary Guard. These letters, which have been appended to Montazeri's memoir, showed that America was secretly arming Iran, and that Ghorbanifar had brought an American delegation, headed by Robert McFarlane, to Tehran. The letters have a historical importance; they set in motion Iran–Contra. They shed a surreal light on the Iranian corner of the stage and the actors creeping about it.

Ghorbanifar deserved his supporting role to the headline stars. He had a handsome beard and a winning propensity for soppiness. He'd adopted the Cold War patter – he said he wanted to turn his homeland away from the Soviet embrace. (This, as he assured his contacts in Tehran, was actuated only by a desire for Iran to win the War.) He had friends in intriguing places; he could supply forged passports, print Saudi riyals and negotiate with the IRA for bombs and booby traps. But Ghorbanifar was a braggart and a poseur, and the Americans were sceptical about his usefulness. They knew that the Revolution had turned Iran from a place that they – superficially, at least – knew well into a puzzle. They were relying for insights into Iran on a man whose polygraph test showed him to be unreliable on every question save name and nationality.

Then, there was the matter of results. By the time of the McFarlane visit, nine months after the first arms shipment, only one American hostage had been released. (The Iranians, realizing that they were on to a good thing,

---

* True or not, such stories have the effect of enlisting the Imam's posthumous support for causes of which he may not have approved. In the case of fallen golden boys like Hashemi, they perpetuate the idea that Khomeini was omniscient; he alone spotted the rotten apples while they were still on the bough. In this vein, it's often said that Khomeini had had doubts about Bazargan from the start. 'If that was so,' Saeed Montazeri asks caustically, 'why did he make him prime minister?'

arranged for another American to be seized.) Even McFarlane's trip, which had as an (unrealistic) aim the immediate release of all the hostages, failed to break the deadlock.* (Two hostages were released after that trip, only for two more hostages to be seized.) By the time of his letters – the longer and more interesting of which is dated 9 July 1986 – Ghorbanifar's importance to the operation had been reduced by the emergence of a Second Channel, a young Iranian official believed to be Rafsanjani's nephew. By then, Ghorbanifar was claiming that his efforts as a middleman had left him seriously out of pocket. This is the purport of his letter of 9 July – a missive of such servility, venality and self-righteousness, it captures the degradation of Revolution.

The letter begins with a scraping bow – with 'salutations and prot-estations of devotion', and an expression of the author's hope that 'you and your esteemed family will enjoy divine grace and health'. By the end of a serpentine opening sentence, Ghorbanifar is already hinting at the misfortunes he has suffered; he pleads that Kangarlu 'pray for my welfare'. He querulously reminds Kangarlu of the latter's ingratitude. This comes in the form of a tangential meander in pursuit of some errant caviar (there's a wartime scarcity) that Kangarlu apparently promised to procure for Ghorbanifar but never delivered. 'Who knows who ate it, and where?' muses the inconsolable Ghorbanifar.

Ghorbanifar's letter continues in this vein for nineteen lachrymose pages.

---

* This trip was to have been the clincher– that's what Ghorbanifar, who arranged it, told the Americans – and it became a surreal adventure. There was no one at Tehran airport to meet the American delegation; according to George Cave, the only Persian-speaker among the visitors, 'the Iranian negotiating team had not believed we would actually come to Tehran'. At length, the Americans, along with one Israeli, were taken off to the dreary Independence Hotel (the former Hilton, with curling carpets and no bar), where they stayed three nights, before leaving empty-handed and bewildered. It was clear that Ghorbanifar had exaggerated the Iranians' willingness to press for the release of the hostages, and exaggerated the Americans' willingness to sell weapons. McFarlane found in his meetings with Rafsanjani, who was in charge of the arms purchase policy, that there was a gulf of understanding and culture between the two sides. 'It may be best for us,' he later reported, 'to try and picture what it would be like if, after a nuclear attack, a surviving tartar became Vice-President; a recent grad student became Secretary of State; and a bookie became the interlocutor for all discourse with foreign countries.'

An aim is discernible through the mist: to sum up the services that he has provided the Islamic Republic in its quest for weapons. (He does this so thoroughly, in fact, that you might suppose the letter had been written for the perusal of a third party.) According to his own account, Ghorbanifar has spent more than two years engaged in this thankless task. He calls on God – in that, if little else, he resembles Oliver North – to bear testimony that he got nothing for his pains save a hole in his wallet. (That lament must have been wearyingly familiar to the American side; despite Ghorbanifar's insistence that the arms transfers were ruining him, they suspected that he was getting steadily richer.) The letter is suffused with hyperbole – as in his fallacious characterization of McFarlane as 'the second figure in the Republican Party, and the person with the most influence over Reagan'.

Perish the thought that Ghorbanifar is performing his patriotic services for the money! He's accepted bad and tardy reward only because he's doing his country a service. According to his letter, Ghorbanifar borrowed the money, all $24 million of it, to pay for the arms that Iran ordered in the run-up to the McFarlane trip. Now, it's forty-six days since the weapons (or rather, some of them) entered Iran along with the American delegation, and the interest is building up. Having received some of the arms, Kangarlu has snapped his wallet shut – he seems to suspect Ghorbanifar, or the Americans, or both, of ripping him off. Ghorbanifar signs off with a flourish. 'I ask of you, dear brother, I entreat you, not because of me but because of God and the prestige of the Islamic Republic and in the name of humanity, to let me know within a few hours how this matter is to be resolved. After that, the responsibility for this is all yours.'

Are those last words – that allusion to 'responsibility' – a threat? By giving copies of his letters to Montazeri's associate, Ghorbanifar's intention was not to provoke scandal but to enlist the ayatollah's intercession on his behalf. But Montazeri had no intention of coming to Ghorbanifar's aid. Montazeri was, as he told me in 2003, appalled that 'behind the backs of parliament and the people, weapons should be purchased directly and indirectly from the Americans or the Israelis in order to continue the War with Iraq, while it was clear and obvious that neither America nor Israel willed Iranian victory; on the contrary, they desired the War's prolongation until the complete elimination of two powerful countries in the region.'

What Montazeri intended to do with the information is harder to say. (Saeed glided past that question when I put it to him.)

The answer may be a kind of pre-emptive blackmail. Hashemi's enemies probably intended to have him arrested well before Montazeri learned about the McFarlane visit. If, as Montazeri says, Hashemi himself disclosed the McFarlane trip to the Lebanese magazine, why didn't that magazine reveal it until after Hashemi's arrest? It's likely that Hashemi regarded his (and Montazeri's) knowledge of the arms sales as a kind of insurance against future misfortunes. Montazeri, by his own admission, casually informed Rafsanjani that he knew. (Rafsanjani asked, 'How did you find out?' Montazeri replied, 'The djinns told me.')

Montazeri's implied threat would have cowed a lesser manipulator, but Rafsanjani knew that the arms transfers had the Imam's support. The Imam, unlike Reagan, wouldn't squirm away from responsibility; his supportive silence would bleach any stain on Rafsanjani. So, rather than put his plans for Hashemi on the backburner, Rafsanjani and his friends in the Intelligence Ministry seem to have accelerated them.

The job against Hashemi was done, says Montazeri, by people who 'hadn't so much as sneezed for the Revolution'. The Intelligence Ministry found evidence to suggest that Hashemi and his friends had defamed President Khamanei. They came across what they said was an arsenal being used by Hashemi. More unlikely, they claimed to have found 'evidence' suggesting that Hashemi had been in Savak's pocket before the Revolution. On 12 October 1986, Hashemi and several of his associates were arrested.

On the Imam's order, a court was set up for the trial of clerics by other clerics behind closed doors – to ensure that dirty turbans weren't washed in public. To give the people proper guidance, Hashemi took part in a notorious TV interview. (Having been tortured beforehand, Montazeri insists.) In this interview, Hashemi 'confessed' that he had 'deviated' from the Imam's line, cooperated with Savak and had a hand in the murder of several people before and after the Revolution. He announced that he'd used Montazeri's secretariat as a 'base for the realization of my own designs', and that his supporters were gaining influence in seminaries under Montazeri's purview. Nowhere did Hashemi accuse his mentor of subscribing to his deviant views. (What were they, and how did they differ from those of his opponents? No one cared. Hashemi wasn't in trouble because of his beliefs or even his methods, but because of his independence.) Montazeri emerged looking like a political greenhorn, smitten with what Hashemi called a 'pure and unsuspecting trust' for his protégé. After further interrogation yielded more detailed 'confessions' concerning Hashemi's involvement in several murders, there remained, in the words of the Intelligence Minister, 'no option but to execute Mehdi Hashemi'.

Did he deserve it? Hashemi's 'gang' were certainly brutal. 'Opposition to Montazeri,' went the slogan in some quarters of Qom, 'is opposition to God.' Hashemi and his boys were prepared to kill in defence of what they regarded as the integrity of the Revolution – and in promotion of themselves. It may be that Montazeri, observing their zeal with paternalistic benevolence, overlooked their excesses. He, while vigorously defending Hashemi from charges of cooperation with Savak, agrees that his secretariat contained few 'saints'. His request that Khomeini pardon Hashemi suggested that there was something to pardon. His letters to the Imam before Hashemi's execution amounted to a craving for Khomeini's indulgence of a fanaticism that they had both stoked.

In any case, many people believe that the Hashemi verdict, and the sentences to execution, exile and imprisonment that were served on dozens of others, were designed to neuter Montazeri politically. With its carefully weighted aspersions on the ayatollah's judgement, the Hashemi interview and the accompanying media propaganda turned Montazeri into damaged goods. Excuses were found to shut or reduce the activities of the seminaries under his purview. His enemies took the opportunity of pinning on the Hashemi gang objectionable actions that, according to Montazeri, the Revolutionary Guard had carried out. And Hashemi's elimination, while presented as a blow for moderation against extremism, didn't end the intimidation and the murders. They were to become the speciality of other groups.

And so, Iran–Contra fell off the front page. Apart from Oliver North, who retained a considerable pull among America's patriotic right, none of the big players on the American side can be said to have won. The three released hostages were winners but their victory was soured by the seizure of three other Americans. It's hard to determine if Ghorbanifar won or lost. Whether or not he eventually recovered his money from Kangarlu, he continued to moan that he had been too patriotic for his own good. On the Iranian side, the Revolutionary Guard won only modestly: the missiles flowed fitfully and briefly. The indisputable winner of the affair was Rafsanjani. Far from being killed politically, he used the crisis to dispose of restraints on his expanding power; he would later take formal control of the War, exert increasing dominance over the Intelligence Ministry and become president.

The truth lost. Although they continued to trust Khomeini's judgement, Iranians no longer expected to hear the truth from their leaders. And the Islamic Republic paid a more specific price. The Mehdi Hashemi affair did not only result in the elimination, from political life, of the people around Montazeri. It alienated the millions of Iranians for whom Montazeri had been an inspiring figure, a brilliant theologian and an example of the ideals – naïve, perhaps, but true – for which the Revolution had been raised.

CHAPTER SIX
Reza Ingilisi

Going east from Execution Square, Tehran's former Tyburn, you skirt the bazaar's southern border. The island in the middle of the square has been surrounded by a rough palisade and half-covered in tarpaulins, a poultice over the gash inflicted by men working on the metro station underneath. There's something furtive about change here, as if the streets are being assaulted but have drawn a veil over proceedings. Hands are pressing on southern Tehran, disfiguring and crushing, reaching in for a fistful of guts. Nowhere do you get the impression that the city's wounds – the potholes and the rusted awning over a declining foundry, a jagged hole where a shop has been demolished – will heal cleanly. Different directions and textures are being imposed, not by design but by indifference and a sick rapacity passed off as progress. The understandable old grid of brick low-rises is being ripped out, and the result isn't diversity but incomprehension.

The bazaar is a square kilometre of social ascents and falls, with rumours and loyalties flying down the lanes. If you go in the early morning to the frontier, Mowlavi Street, you find porters pushing metal carts to the southern entrances. There are policemen on street corners with five-day beards, bottle-green uniforms and monthly pittances; they prowl for motorists to fine and for the grace of a small bribe. Every so often they speed ostentatiously down bus lanes in their newly acquired Mercedes-Benz sedans. The shutters are rattling up on shops selling boxes folded flat and cardboard egg-racks, and the bazaar beggars are coming out. (There's one little boy whose face when I first saw him was so stricken with agony and desperation, I hurriedly gave him a big note. The next time, he was wearing the same expression, and the next; gradually, I began to realize that he adopted the expression every morning before starting work, like a lawyer putting on his tie.) Vile-smelling addicts shuffle from the parks to breakfast in a cheap café near the bazaar, and thence to their next hit. Everywhere is grey, brown and black – colours that, if you stay long enough in Tehran, start to become mutely reassuring.

Akbar Barjasteh's house of strength is about three hundred metres down Mowlavi, in a narrow lane. You might stumble upon it during a seance – what you're about to witness cannot be called a performance, still less a bout – in which case you will feel like an extra in some theatre of the absurd. A workout seems to be in progress – of freemasons, perhaps, in a Turkish bath. Standing in an octagonal pit sunk a metre into the floor of a room the size of two squash courts with a high roof and skylights, a dozen or so middle-aged men wield two heavy wooden clubs apiece, bringing them alternately over their shoulder and back in front of their chests, like the pistons of a locomotive. The men wear red loincloths, and some have stripped to the waist to expose lavish tattooing. Only in professional darts can men with bellies as commodious as these call themselves sportsmen; they would surely collapse if they had to run a hundred metres.

Seeking the racket that accompanies the lunacy, you look left to where a half-naked man sits in a kind of pulpit, his handsome face framed by a glittery arch of Moorish inspiration. This is Akbar, owner of the house of strength and the 'guide', or choreographer and percussionist. His father set up the Mowlavi house of strength three-quarters of a century ago. Akbar is bald and he strikes a large drum with bandaged thumbs, smashing his hand occasionally against three bells of different sizes, suspended inside one another. He calls on the sportsmen to shout a *salavat*, an entreaty in Arabic for divine compassion on the Prophet and his family. Then he starts singing, in stentorian tenor, verses from an epic dating from the turn of the second millennium:

May God Who made the sun and moon, Who holds
The World and Who bestows the crown and throne,
Sustain your heart in happiness forever
And keep it free from pain and all misfortune,
Rejoicing always in your victories,
Your glory, the casque of greatness and the crown,
Of your nobility. You led our troops out,
Eager for war; fortune, skill and righteousness
Are yours and, though the smell of mother's milk
Still scents your mouth, your strength in battle snapped
The whippings on your bow. Now may your body
Keep its prowess always, and may your heart
Attain to its desires.

It's the paean of Kavus the king, to Siyavash his son, lion-hearted warrior
and victim of the intrigues of royal courts. The account of Siyavash, an
episode from Ferdowsi's mythology of pre-Islamic Iran, The Book of Kings,
is about the conflict between conscience and service, about resistance to
tyranny and also to lust – the Phaedra-like lust of a stepmother, no less –
and about heroic defeat at the hands of schemers and cowards. Siyavash,
like a pre-Islamic Hossein, submits willingly only to the will of God.
Through Akbar he sings:

All praise be to God, sole source
Of victory and fortune, Lord of the sun
And of the circling moon, Who raises up
Men's crowns and thrones and diadems – whomsoever
He wishes He exalts, while to another
He brings lamentation and adversity –
In whose commands there's neither 'why' nor 'wherefore',
To whom as guide our wisdom must submit.

I've read a book in Persian about houses of strength, written by a man
called Tehranchi. In this book, which Akbar lent me, Tehranchi suggests
that the old houses of strength were nodes of 'resistance against the Arab

and Mongol oppressors'. Starting around the time of the Arab conquest, and becoming widespread by the Mongol desecrations six centuries later, Tehranchi says that patriotic Persians got together in secret to practise the arts of war, or 'traditional sports', for use against the invader. (The wooden clubs, for instance, correspond to Persian maces that were brought down on alien heads.) Some historians, however, have used similarities in plan between houses of strength and Zoroastrian fire temples to posit pre-Islamic origins for these sports. Few modern exponents are comfortable with such theories; although they consider themselves within the tradition of a martial defence of the homeland, most of them also regard the traditional sports as an expression of their Islamic faith. Whatever the truth, the spirit of Zoroastrianism, if it ever existed in houses of strength, has been extinguished. They're temples for Iran and Islam.

On entering the house of strength I take off my shoes. I salute Akbar, ensconced in his pulpit holding a glass of tea. Then, I go around the room greeting the other sportsmen, making sure I address the older and more experienced ones first. I say, '*Salaam*, Mr So-and-so', and he and I kiss on the cheeks three times. He says, 'Ya! Ali!' and I say, 'How are you?' He replies, 'Thanks be to God!' I say, 'I am your devoted servant!' or 'I am your sacrifice!' From his sedentary position, he makes a slight bow and puts his right hand over his heart, which means I can go and change into my loincloth.

I go to Mr Soroush, who looks sixty-five but has youthful eyes, and then to Mr Haji Seyyed Javad, saturnine, benevolent; he's calling on the Imam Ali to bless the souls of those who have departed this world. There's a third sporting elder I've not met before, Mr Habashizadeh; I admire his silky white hair. Along with little Mr Pakizehtan – who used to be a champion diver, and whose endeavours are now confined to token club wielding (he rests the clubs on his shoulders and gives them a wiggle) – these are the old guard. They were sportsmen well before the Revolution, and they wear the twill jackets and disciplined moustaches (and shaven cheeks) to prove it.

They call me Reza. Reza's the eighth Imam, the only one buried in Iran, and it's the ceremonial name that Bita chose for me – without due consultation – when we married. Somehow, in south Tehran and Qom and other places where 'Christopher' is hard – or even distasteful – to pronounce, I've become Reza Ingilisi, or English Reza. Once, before I met a mullah whom I knew to be anti-British, I decided to identity myself with the paper I was writing for, and introduced myself as Reza from *The Economist*. The mullah looked puzzled and, when a second mullah came into the room, presented me as 'Reza Communist'. I reverted to Reza Ingilisi.

After the old guard, I pass on to the younger generation, who are distinguished (besides their age – late thirties to early fifties) by their revolutionary appearance. Take Mr Tirafkan, the hosier. Mr Tirafkan wears dull clothes and trims his beard rather than shaves it. There's a second Mr Reza: he has penetrating eyes and stubble on his head; from the vigour of his sport, I'd guess that he let out a lot of aggression in 1979. Numerically, the two groups must be about equal and they seem to accept each other. The old guard were too set in their ways to adapt much to the revolutionary orthodoxy. When they come to the house of strength, the younger lads leave their pushy egalitarianism at the door. They take their place, perhaps with relief, in the caste system of sport.

Soon enough, across the banter and jokes, you pick it up; there's a hierarchy in the house of strength – unobtrusive, even intuitive, but essential. It favours those sportsmen who, besides being skilled and experienced, manifest the piety and chivalry that the house of strength demands. (I use the word 'manifest' advisedly, for the house of strength, while demanding impeccable behaviour in public, is tolerant of private lapse.) The guide reinforces this hierarchy. If he welcomes you verbally when you come in the door, you know you're on the first rung of the ladder. If he calls for a *salavat*, you've ascended further. If he summons a *salavat* and strikes his drum at the same time, you're on the third rung; together, a *salavat*, a drum roll and a ring of the bell signify further advance. And so on. If you get a bell ring for your every significant action – your entrance into the house of strength, your descent into the pit and at peak moments during the sports – it means that you're one of Iran's few champions and the kind of athlete whom everyone should emulate. I find the ascending honours interesting but academic. I get no *salavat*, no bell ringing. My entry is greeted in silence.

Having put on my loincloth – knotting it at the front and tucking the hanging back end between my legs and into my waist – I warily approach two thick wooden slabs leaning against the wall. They're arch-shaped, almost a metre long and more than half as wide, with square holes in the middle spanned by handles. Their modern-day application is limited; the slabs correspond to the shields that the Persians might have used for defence against enemy spears. (Lifting them, Akbar implausibly claims, is practice for the essential business of building 'bridges' of men holding their shields horizontally across pits dug by the enemy.) I lie down on my back, stretch my arms out behind me and grip the handles. The idea is to raise the shields vertically above you, allow them to bow to each other so that their tips almost touch, and then bring them to a horizontal position over your chest, before starting again. It's customary, while doing this, to run through the names of the twelve Imams – an Imam for each lift. I can't remember the names of the twelve Imams, which is just as well; my strength fails me on lift five, and one of the shields topples onto my chest, pinioning me to the floor until Mr Tirafkan comes to my rescue. He removes the shield and I get to my feet, my prestige bruised.

The other athletes are going down into the pit; the act of descending accentuates their humility before God. As they drop down, the sportsmen reverentially touch the floor of the pit before dragging their fingers over their lips and forehead. For a few seconds, they bunch at the end of the pit that's furthest from the door – the 'bottom' end, where the inferior sportsmen are meant to stand. It's finely calibrated false modesty; the old guard are insisting that the young lads take their places at the 'top' of the pit. 'I beg of you, Mr Haji Seyyed Javad', Mr Tirafkan dutifully remonstrates, 'please go up!' For a few seconds, Mr Haji Seyyed Javad and the others resist, and there is much good-natured jostling. Eventually, the old guard succumbs.

The hierarchy confirmed, the bottom of the pit thins out to the extent that I can leap down. (When I first went to the house of strength, I'd stay on the lip overlooking the pit, protesting my unsuitability for the privilege of debasing myself before God and my peers. I did this because Mr Tirafkan said I should, and at first I found all the contrapuntal insincerity pretty confusing, before getting the hang of it, and eventually coming to enjoy it.) Mr Habashizadeh has bowed to demands that he take the role of a kind of object of emulation – the athlete standing in the middle of the pit, whose movements are to be followed by the others. He is now delivering to the rest of the sportsmen – who have gathered up their clubs – a homily to a Mr Habibi of Sepideh Street. Yesterday, the white-haired man tells us, 'God saw fit to take Mr Habibi to his side, after a long struggle against illness. He was one of the pure acolytes of our community, whose love for the Imams he carried into his trade of making kebabs, which he did for generations with a pride and joy that invigorated the bodies and souls of the good people of the locality' –

'Especially the coppersmiths in the bazaar!' Mr Hajji Seyyed Javad interjects. 'Their excellent work was proportionate in no small measure to the excellence of his mutton.'

Mr Habashizadeh frowns at the interruption. 'Now, to the spirit of that God-fearer and devotee of Islam – may God forgive him his sins and grant him an eternity of peace in the sun-filled uplands of heaven! – to his widow and his two daughters of marriageable age, to his three brothers, one of whom was a sportsman in this very house until he was gassed in the Inflicted War against the Satan in Baghdad – to them, willing the gift of fortitude in their time of enforced solitude, *salavat!*'

There's a thunderous *salavat*, the drum rolls and the sports begin.

❖ ❖ ❖

If you ask the sportsmen they'll tell you that the values inculcated in the houses of strength – gallantry, piety and physical prowess – are the values that once ruled south Tehran. For a while, I saw nothing controversial in that – until, that is, I told my father-in-law I'd taken up traditional sports and he looked at me very oddly. I'd been expecting an indulgent chuckle; in fact, his expression of puzzlement gave way to one of concern, though he restricted himself to advising me not to believe everything they told me. (He also recommended a good swimming club close to home.) As warnings go, this was tantalizingly opaque; it was further reason to carry on going to the house of strength.

Now, more than a year on, I know the culture of south Tehran, and the house of strength, a bit better. My father-in-law's disapproval stemmed partly, I realize, from the old gossip about pot-bellied mace-wielders and young athletes. (I know some artists who have treated houses of strength in a homoerotic way, though some of them are predisposed to such inferences.) Class probably played a part. Bita's father may have been upset that his imported son-in-law, a well-brought-up chap who claimed to have attended the University of Cambridge, was enjoying the company of the lower middle orders. (The house of strength may once have been a proletarian place. It no longer is. You need leisure, the hallmark of the middle class, to use it.) But there's a more historical reason for his disfavour. Back in the 1950s and 1960s, a heyday of traditional sports, the house of strength was the preserve of the Tehran mob.

Don't think of competing gangs, killing and intimidating for control over narcotics, liquor and whores. You never hear the people of south Tehran, even the older ones, talk about a mafia in the mould of Palermo, New York or Bombay; Tehran didn't have a mafia, but dozens of 'thick-necks' – an idiom that evokes with poetic precision the otiose exercise of their power. One thing they had in common with the godfathers of New York and the gangsters of London's East End: they were known for their fear of God and the love of their mothers (or perhaps the other way round). But they were gentler; their weapon of choice was the knife, not the Colt. While they handled, by the standards of the times, considerable sums of money, their goal was not to monopolize business. Much of the time they contented themselves with receiving 'contributions' from those who did. Such contributions rarely went towards buying villas or limousines or factories. They were often redistributed, which made the thick-neck the most beloved and philanthropic of crooks. Nowadays, some people maintain that the thick-necks filled the vacuum of authority that emerged after Reza Shah's abdication; without them, they argue, society would have become lawless and immoral. It might be more realistic to say that the thick-necks had a monopoly on lawlessness and immorality.

They patronized wrestlers who used to fight in the pit. Then, after the Second World War, the authorities modernized wrestling, taking it out of the pit and putting it on mats in sports halls, a revolution that generated brilliant fighters who won gold medals at the Olympics, and which changed irrevocably the house of strength. Not that these developments undermined the thick-necks' allegiance to traditional sports. For them and their followers, the house of strength remained a barracks, clubhouse and prayer room. It was here that warnings were issued to young rowdies, and community and religious values were instilled. (The whoring, boozing and knifing went on elsewhere, mainly in a self-contained, sinful heaven called the New Town, which got bulldozed after the Revolution. On home territory, in sight of the women and children, the thick-necks demanded puritanical levels of moral austerity.)* Each year on the anniversary of the death of the Imam Hossein, athletes from the houses of strength were the nucleus for vast self-flagellating congregations, organized by thick-necks in competition with each other, that processed through the streets. (The strongest sportsman, naturally, prided himself on carrying the heaviest standard.)

In Akbar's house of strength you find references to all of this. Look past the side wall – a photographic roster of departed athletes, some of them martyred in the War – to the far end of the room. Attached to the wall is an iron standard that is used every year in the mourning procession for Hossein. Arranged around it are dozens of photographs, most of them black and white. In one old portrait, two hairy men with parabolic moustaches and baroque plus fours glare at the camera while gripping each other's belt. Behind them, a third athlete disports himself with a heavy iron contraption – it's part of the traditional house of strength pantheon and corresponds to the longbow of old. There's a photograph of that incomparable wrestler and gentleman, Ghollam-Hossein Takhti, in the company of Butcher Goli, the most senior of the sportsmen who continue to frequent Akbar's house of strength. (When he visits, Butcher Goli invariably wears a navy-blue suit over a white shirt and an avuncular smile. After languidly accomplishing his sport, he sits down at the far side of the room

---

* I'm reminded of an extraordinary story, related to me by Akbar, and which everyone in the house of strength insists has not been embroidered. The story concerns a thick-neck whose associate was going on the pilgrimage to Mecca; as was the custom, the pilgrim entrusted his wife and children to the thick-neck. That meant that they stayed in the thick-neck's house. One morning, forgetting that he had guests, the thick-neck walked into their room, and saw the wife of his friend naked. The thick-neck was distraught. He cried out, 'As God is my witness, I did not sin, but my eyes did.' He took out his knife and blinded himself in remorse.

and smokes the hookah that one of Akbar's sons brings him.) There's a shot of Mr Pakizehtan in the days of his triumphant freefall, limbering up for a champion dive.

The most evocative photograph is high up the wall. It shows Akbar's father – he's bald and handsome, like his son, but his eyes have a resolve that Akbar's lack – and a dozen athletes. They all wear plus fours and proud smiles; it's the Shah's birthday, and they've been invited, along with the representatives of some forty other houses of strength, to participate in a display of traditional sports for the King of Kings. One year, Akbar took part. Some four hundred traditional sportsmen emerged from different entrances onto the turf at the Amjadiyeh Stadium, accompanied by ten drummers playing in unison. The Shah's favourite thick-neck, Brainless Shaban, led the sports.

So far, so conventional; these are analogous to team photographs on the rugby club wall. Then, you come to a portrait that seems out of place among the sporting poses. It's an important image, you realize; it's been privileged with the sort of gilt frame you see around a lurid landscape of the Perso-Flanders school. (You'll find such a landscape in an affluent north Tehran lounge, along with an imitation Louis XV occasional table and a vase full of synthetic roses with plastic dew.) Five men and two of their sons stand at an oblique angle to the camera, in front of a heavy commode. At first, your eye is caught by the whale-like build of one of the men, a colossus with a white jacket and a Hitler moustache. Then, you realize that, despite his prominence, he doesn't occupy the middle of the photograph. That honour belongs to a smaller man who looks on pensively, past the camera. It's the thickest thick-neck: Teyyeb.

He's the strongman for which south Tehran, with its weakening social fibres and flight to addiction, is nostalgic. He grew up in and around the fruiterers' wholesale market, where his dad had a stall. Quick on his feet, quick with his hands; he was appointed by an important trader, Arbab Zeyn Al-Abedin, to man one of the market gates, levying a tax on every fruit-laden truck that entered the city. (The fruit arrived on camel back, before being transferred onto trucks.) The tax revenue went to traders like al-Abedin, and the municipality got its cut. It can't have been easy work; the retailers were rough fellows, and crafty, too. But Teyyeb, supported by a group of heavies, turned out to be an effective collector – so effective, in fact, that Al-Abedin rewarded him with a stall of his own. Teyyeb was able to buy one of the scales in the market, which increased his influence as a regulator. People started asking him to mediate in disputes. He became an unofficial spokesman for the market. He had good links in the houses of strength. He'd proved his spurs in jail. On one occasion, he distinguished himself in a famous prison brawl between Tehranis and some Turks from the province of Azerbaijan, damaging Turks with a deftly wielded piece of wood.

Such are the bare facts, but they're inadequate to explain why Teyyeb's portrait adorns Akbar's and other houses of strength, as well as shops and cafés across south Tehran. It may be that the current nostalgia for Teyyeb has something to do with the worldliness of Persia and its relish for the morally ambiguous. (The Islamic Republic might be seen as a joy-killing riposte.) Teyyeb was ostentatiously, hyperbolically pious, but he didn't claim moral perfection. He had a gold coin attached to his shirt as a kind of brooch, on which had been inscribed, 'Ya! Great Ali!' But he never made the pilgrimage to Mecca, for that would have implied a repentance that he couldn't entertain. He lapsed in the New Town, that Shangri-la of immorality, but desisted during Shi'ism's three months of solemn religious mourning. Think of Robin Hood with a knuckleduster. The people of south Tehran couldn't help feeling that the good outweighed the bad; space was surely made for him in heaven.

Teyyeb's business was handling money, passing it on, trickling it into crevasses of want. He helped the poor, promoted religion, kept the peace. The line between extortion and charity collection was unclear. But even now the people of south Tehran – the authentic south Tehranis, I mean, not the uncouth yokels who've flooded in from the provinces – recount stories that illustrate Teyyeb's love for the people and God. These tales are too numerous and diverse not to contain some truth.

Take the story of a penniless young fiancé for whom Teyyeb collected enough money from the market traders to buy a matrimonial home. Or accounts of Teyyeb doling out money to released jailbirds, to help them get back on their feet. Teyyeb fed five thousand people to mark the anniversary of the Imam Hossein's death; he bought winter coal for four hundred poor families. Such was his reputation for philanthropy, old women would ask for him at his scales in the market long after his death. He had a bumbling flamboyance; he once turned up at a wedding with a bouquet so big it wouldn't get in the door. In the name of the Imam Hossein, he and a second thick-neck, Icy Ramazan, organized the greatest parades of flagellants and iron standards in all Tehran; their separate processions would meet and then be joined by others, becoming a vast millipede of mourning. (Teyyeb would take his place at the back of the parades, to underscore his humility.) He was a benefactor in love, too. He took as a second wife an 'outdoors girl', in the parlance, a prostitute whom he is credited with rescuing – from a lifetime of social odium and an eternity in hell.

I'm thinking of my favourite Teyyeb portrait. He stands in the corner of a white-walled room, wearing a Homburg, scarf and thick overcoat. He seems almost isolated in the whiteness of his corner, with his too-long trousers and a moustache like a fly on his upper lip. There's something anti-star about him – appropriate to a culture that celebrates self-deprecation. (No one talks about Teyyeb's wit, his sparkling personality.) You'd never think that this nondescript fellow had a decisive effect on Iranian history.

❊   ❊   ❊

One morning in the winter, after sport, I walked about a kilometre east along Mowlavi before turning south and then right into a lane with low houses and a few shops. A little way down the lane, I came to a single-storey building with a big window. Through the window I could see an old man; he looked so brittle and insubstantial he could have been a drift of leaves. It was Nasrullah, Teyyeb's buddy, and he was dozing in front of a gas fire. I opened the door and one of his eyes opened; the other was a lurid socket. He bade me take a seat. (At least, that's what I understood from his congenial gestures and tone; toothless gums, combined with a general decrepitude and smoker's rasp, complicated the task of understanding him.) He was wearing a copiously stained three-piece gingham suit of antiquated cut that was garlanded with two fob watches.

He was known semi-accurately as Blind Nasrullah, and his profession was to rent out and (formerly) to lift heavy objects that have to do with important and symbolic occasions in your life. (I'm puzzled by the emphasis placed in Iran on lifting things – the heavier the better – for obscure reward.) He started his career as a tray-lifter – that's to say, the bearer (on his head) of one of the enormous trays that would form a ceremonial procession of the bride's effects, from her parents' to her husband's house, on the eve of a wedding. (Such processions might have dozens of trays, depending on the affluence of the bride's family. One would be laden with the heavy mirror, electric lights and other symbolic goodies – pomegranates and quince, eggs painted silver and gold, a clod sprouting green shoots – that are central to the wedding ceremony. The others would bear the bride's suitcases and household appliances.) Skirting formal education, Nasrullah graduated to heavier items: the 'forty lamps' – a generator-powered chandelier that was placed outside the bridal home – and finally to the *hejleh*, a vast iron, tent-like canopy, surmounted by light bulbs, feathers and other baubles. (This would be placed in the new home, with a bed or mattress inside, as a bridal suite.)

With the arrival of heavy appliances like fridges and washing machines, the tray-lifters were replaced by vans and the wedding processions died. Nowadays, *hejleh*s are associated more with memorial services than with romance. (The services happen on the third, seventh and the fortieth day after death.) This new association came about during the War, when it was said that unmarried martyrs had contracted a marriage with God. Rather than be borne aloft in a procession, you often find the *hejleh*s resting in the street, pious announcements of a death in the house opposite or signposts to a mosque where a service is to be held. (They are often adorned with photographs of the deceased.) You also find them outside places where men are gathering to beat their chests and weep for Hossein. Nasrullah has stopped lugging them for a living but he still rents out his *hejleh*s; he has half a dozen of them. Every day he sits in his office, waiting for customers.

He stopped swinging the clubs a few years back. (I once accompanied him to a house of strength – not Akbar's – into which he was reverentially ushered as a presiding muse.) He's stronger than he looks, as I discovered during the tour he gave me of his office, which he spent squeezing my shoulders in a disconcerting and painful manner. Lately, he's achieved modest celebrity, giving long interviews for a new book, by a man called Sina Mirzai, about Teyyeb. On one visit, I found him being cantankerous with a handful of students – aspiring journalists, perhaps, or filmmakers – who had come down from north Tehran to interview him. He may have detected in these young men a lack of patience, a lack of respect; rather than take the trouble to make friends with him, stroking his vanity and bringing him cake, they'd turned up empty-handed, expecting instant gratification. Citing lack of time, he refused their entreaties. If there's anything Nasrullah has in abundance, it's time.

He's turned his office, which is about the size of an industrial container, into a remarkable photographic museum. For those, like me, who butter him up and bring him cake, the guided tour starts with a black and white portrait of a slick young man in suit and tie. (That's Nasrullah shortly before smallpox cost him an eye when he was twenty-five.) Next comes a one-eyed Nasrullah wearing the egalitarian white garment that Muslims put on for the pilgrimage to Mecca. On the same wall there's a fine shot, taken about a decade ago. It shows Nasrullah naked from the waist up, showing off an impressive acreage of tattoos, pale blue from the juice of crushed cress. Mermaids and Imams are the main subjects, and they look nice against the studio backdrop, which resembles wallpaper on a child's bedroom wall and is covered in aeroplanes. The final Nasrullah depicts him in the Amjadiyeh Stadium on the Shah's birthday, walking with an enormous *hejleh* on his head. Imagine thousands of feet stamping in appreciation.

The walls of Nasrullah's office show an underworld that never really went under. Its home was south Tehran – among the low-rise tenements and houses of strength, and the by-lanes of the bazaar – but its influence extended south beyond Shah Abdolazim's shrine, and north as far as the court. For a few years in the 1940s, following the promulgation of a law prescribing exile for violent thick-necks, it moved to the Persian Gulf port of Bandar Abbas. According to Nasrullah, one hundred and thirty-two 'ruffians' were sent there – which must have been as much a shock for the inhabitants as it was for the exiles. (Teyyeb and Nasrullah were among the latter. Teyyeb had been fingered for involvement in the murder of a certain Cocky Muhammad. Unjustly, protests Nasrullah; Cocky Muhammad, he says with engaging earnestness, knifed himself to death.)

Their names are more memorable than their puffy, plebeian faces. There's Teyyeb's friend and rival, Icy Ramazan (he was too prudish to look at a woman – unless, of course, he was lying on top of her in the New Town); Devilish Ali (named for his mischievousness); Cockroach Asghar; Ali Panther; Skull-Cooker Mehdi (his restaurant specialized in sheep's heads and intestines); Mahmud Camel-Lip; the Seven Bald Men (not all bald, by any means, but a formidably ugly gang of brothers); Greedy Guts Haj Kazem; Early Bird Ja'far. Special mention should be made of Mad Mustafa, who made his living extorting and running a gambling den. (He and Teyyeb fought on several occasions.) He earned his moniker during World War II, when Germany was popular and he was drunk. Wrenching down the imperial symbol that was hanging outside a local police station, he is said to have shouted: 'There should be a picture of a lion here, and I am that lion. Long Live Germany and Long Live Mad Mustafa!' (He later repented, changed his ways and went to Mecca.)

Teyyeb, Nasrullah's friend and patron for thirty-five years, naturally occupies the most space. Nasrullah observed and participated in the giving, fighting, roistering. Here's Teyyeb stout from a ballast of kebab; here, wearing an unlikely white linen suit dimly inspired by the Chicago mob; holding his potato-like son; sitting at a table with other thick-necks (between the knife fights there was lots of uneasy fraternizing). Teyyeb was the best thing that happened to Nasrullah, the source of his best reminiscences: 'We were drinking booze ... and there was an architect who was a little bit rude and we taught him a lesson and they took him to hospital and after that we became friends ...'

Most of them went to jail for short periods, and the reason for the leniency can also be found in Nasrullah's photographs. Every so often, you come across a patrician in a well-cut suit. He's a representative of the court, or in with the government. He'd have had a word with the local police chief to ensure that a favoured thick-neck could pursue his interests without harassment. (One official is said to have arranged for Teyyeb's exile in Bandar Abbas to be cut short.) One picture shows an intriguing event in the fruit and vegetable market. The market has been encircled by chairs that are occupied by local notables, including Teyyeb. The seats of honour – armchairs that have been put on a carpet – are occupied by two officials and a uniformed army officer. They lounge like Edwardians at a durbar, apparently waiting for a display or a performance, helping themselves from a table decked with fruit.

It's the reciprocal fascination of the elite and the mob. When Icy Ramazan and Teyyeb fought over a woman, the future head of Savak brokered a reconciliation. Think of the famous and ultimately fatal dispute between Teyyeb and Colonel Nassiri, the commander of the royal guards; that enmity started because of a lap-girl called Tamara. (Thick-necks and senior officials frequented the same gaming clubs and brothels.) For a while, at least, the Shah seems to have liked his thick-necks. They gave him the feeling – fallacious, but reassuring – that he and the common man understood each other. Teyyeb and Brainless Shaban would greet him on his return from foreign trips, sacrificing sheep and even cows on the tarmac, throwing carpets at his feet. The Shah was delighted when, to congratulate him on the birth of the long-awaited crown prince, Teyyeb and his pals ceremoniously lifted up the royal limousine – with the sovereign in it. (Lifting again.)

The thick-necks were useful because they and their followers supported the monarchy. (So long, that is, as the monarch didn't interfere with their religion. In that respect, until he tangled with Khomeini, the people of south Tehran preferred the Shah to his father, Reza Shah.) They shared the court's abhorrence of Communism and had good relations with conservative clerics. In 1953, this alliance engineered its supreme achievement. In that year, nudged by the CIA, a government fell.

The government was headed by Muhammad Mossadegh, an eccentric but brilliant nationalist who had incurred the wrath of the British by nationalizing the Anglo-Iranian Oil Company in 1951. Although Mossadegh was admired by millions of Iranians, he was soon opposed by a formidable coalition. The Shah feared (incorrectly, it turned out) that Mossadegh wanted to oust him and install a republic. The Americans feared (again, incorrectly) that he was a cat's-paw for Iranian Communists; with him in power, they feared that the Soviets would suck Iran into their sphere of influence. Conservative clerics feared (correctly, perhaps) that Mossadegh would steer the country towards secular democracy. Together, the anti-Mossadegh coalition controlled the press, exercised a strong hold over sections of the security forces and carried CIA dollars to buy support. Twice before Mossadegh was toppled they came together to try and kill or arrest him; after the second (failed) attempt, the Shah fled to Rome. (Even in youth, the King of Kings was a coward.)

A few days later, on 18 August 1953, Teyyeb sent word to his closest associates to gather in the fruit and vegetable market. The following morning, some three hundred came; they carried knives and other weapons. Teyyeb doled out money that he had received from a trio of banking brothers who were acting as a conduit for CIA money. Swollen by more local lads, the crowd set off from the market, raising anti-Mossadegh slogans, beating up people wearing white shirts (a fashion associated with Communists), and forcing passing cars to hoot in support of the Shah. Along the route, Teyyeb's band joined forces with a second group, led by Icy Ramazan. As they marched north, the accretions continued. The most picturesque group came from the New Town and was composed of celebrated prostitutes of the day. Notwithstanding the odd diversion to trash offices associated with the Communists, the mob's target was Mossadegh's house in Palace Street.

Surprise contributed to the coup's success; those members of the security forces who had not already been bought or pacified by the plotters may have confused the agitation with other demonstrations, some of them opposed to the shah and instigated by CIA-funded *agents provocateurs*, that had taken place over the previous few days. No one interfered as the demonstrators marched to Palace Street; by late afternoon, Mossadegh's house was encircled by the mob and rebelling army units. During the ensuing gun battle, the prime minister fled over the wall into a neighbouring house. (He was arrested the following day.) Teyyeb dallied long enough for army officers, right-wing journalists and other pro-Shah worthies to come and congratulate him. Then Mossadegh's house was set on fire.

The actions of Teyyeb and the others allowed the coup makers to portray the events of 19 August as an expression of popular will – rather than a CIA-sponsored plot. A few days later, Mossadegh's successor as prime minister hosted a reception in honour of Teyyeb, Icy Ramazan and other mobsters. Teyyeb and Icy Ramazan were presented with pieces of prime agricultural land. Teyyeb was further rewarded with a tax-free monopoly on the import of bananas into the capital, and an ingenious piece of equipment that turned his basement store into a hothouse. The Shah awarded Teyyeb first refusal for a term of five seasons on fruit from the best royal orchards.

Within a few years, the affair soured, partly because of another's jealousy. In 1959, Queen Farah gave birth to the crown prince in a south Tehran hospital. For the townspeople it was a singular honour; in those days, chic Iranians usually gave birth in London or Paris. Teyyeb, who got his boys to stand as a guard of honour around the hospital, was infuriated when Nassiri brought in extra policemen to form a *cordon sanitaire*. The implication that his own boys couldn't keep order, or that they constituted a threat to the royal mother and child, was very insulting. When the Shah came to meet his heir, Nassiri tried physically to prevent Teyyeb from approaching the Shah bearing a ceremonial silver brazier with smoking wild rue seeds to ward off the evil eye. Having elbowed Nassiri out of the way, Teyyeb took up the matter of the *cordon sanitaire*; the Shah ordered Nassiri to withdraw his men. For the colonel, the scuffle and the Shah's rebuke were a great humiliation and are widely supposed to have ignited his desire for revenge.

Nassiri got the local police chief to make life more difficult for Teyyeb. His bananas started turning up late, which lost him business and standing among his creditors. He was briefly detained for striking a police officer who had denied him a former privilege. Debts led to a skirmish in which he was knifed and badly injured. The extent of his fall from favour was illustrated by the fact of his arrest upon release from hospital, on the trifle of a bounced cheque. (A large one, admittedly.) For a man who'd prided himself on his rapport with the sovereign, it must have been a terrible letdown.

It coincided with changes in Iranian politics. With the death of Ayatollah Burujerdi in 1961, and Khomeini's rise to prominence, the Shah's alliance with the senior clergy was at an end. In the past, the citizens of south Tehran had been reassured by this alliance; it allowed them to express their native monarchism and religious conservatism. Now, in the Tehran bazaar, pro-Khomeini groups were organizing a new politics. They opposed the Shah's modernizing schemes – these were emancipatory of women, and tainted with un-Islamic notions of land redistribution. In order to make their voices heard, they needed powerful men to counter pro-Shah thick-necks like Brainless Shaban. Who better than Teyyeb? Teyyeb had been persuaded by allies of Mossadegh to desist from breaking up opposition meetings. But it was the Khomeinists who cultivated him most effectively.

In March 1963, the Shah committed an act of folly. His thugs attacked the Feyzieh seminary in Qom, hurling three seminarians to their deaths from the roof and making a bonfire out of turbans and gowns. It was an insult against religion, against God. Three months later, on the anniversary of the martyrdom of Imam Hossein, massive crowds gathered around the country, and the event was turned into a commemoration of the dead seminarians. In the Feyzieh, Khomeini warned:

> Let me give you some advice, Mr Shah! Dear Mr Shah, I advise you to desist . . . I don't want the people to offer up thanks if your (foreign) master should decide one day that you should leave. I don't want you to become like your father . . . During World War Two, the Soviet Union, Britain and America invaded Iran and occupied our country. The property of the people was exposed and their honour imperilled. But God knows, everyone was happy because [Reza Shah] the Pahlavi had gone . . . Don't you know that, if one day some uproar occurs and the tables are turned, none of those people around you will be your friends?

Teyyeb's indignation at the Shah's contempt for the clergy may have coalesced with his own sense of rejection. (And, perhaps, with the money that pro-Khomeini *bazaaris* were giving him.) On the day of Khomeini's speech, he and another thick-neck provided protection for tens of thousands of people as they marched through Tehran to protest at the murders in the Feyzieh. When Khomeini was arrested two days later, Teyyeb's boys were called out by the *bazaaris*; they led mobs that ran amok, attacking government property and shouting anti-Shah slogans in the face of murderous firing by the security forces. (Several hundred demonstrators were killed.) Teyyeb didn't take part, but many from the fruit and vegetable market did, and it was common knowledge that Teyyeb had endorsed a decision to shut down the market in protest at Khomeini's arrest.

A few days later, Teyyeb was arrested for organizing the riots. Nassiri had been named as military governor of Tehran and he needed a scapegoat. Nassiri tried to get Teyyeb to admit that he'd been paid by Khomeini to organize the riots. Such an admission would not only have damaged Khomeini's prestige; since the money was alleged to have come from Islamic militants in Egypt, it would have opened the way for Khomeini's trial on the capital charge of treason.

Pride must have contributed to Teyyeb's decision to refuse; he was determined not to succumb to Nassiri in this, their final duel. But his trial – and his refusal to denounce Khomeini – acquired a higher significance. It became a vehicle for personal redemption, a way to atone for a lifetime's sinning. By the time of his death by firing squad a few months later, Teyyeb was being compared to Hor – the same Hor who defected from Shemr to Hossein before the battle at Karbala. Not bad for a south Tehran mobster.

All the same, Teyyeb's execution seems an odd, tangential ending. South Tehran has an exaggerated respect for *seyyed*s. (The innate monarchism.) But Teyyeb wasn't much interested in politics, beyond beating up Communists. He was no revolutionary. (In this, he typified people of his background, before reckless development and dictatorship stirred their doughy conservatism.) The last close-up photograph I have seen of Teyyeb, cuffed and being led to death, shows a bewildered man, unready for his place in history.

After 1963, the loose associations of thick-necks and their boys were gradually superseded – by parties and groups, caucuses and cells. Activist *bazaaris* and radical clerics would become the agents of tumult and change, while the houses of strength stayed on the sidelines. As for Teyyeb, that second Hor – he was forgotten, a stepping-stone.

After 1979, Khomeini barely referred to Teyyeb. The Revolution favoured Islamic vigilantes; they regarded the thick-necks as impure. In the houses of strength they took down portraits of the Shah and put up portraits of Khomeini. Dozens closed. (That was before the current assault – of gyms, pool halls and coffee shops.) No more than forty survive today.

Nodding in his office, Nasrullah wants nothing more than to exonerate the King of Kings. At the time of Teyyeb's arrest, he assures me, the Shah was away. He was ill. His Majesty had nothing to do with the execution; it was Nassiri's fault. This contradicts accounts that the Shah intervened to make sure that Teyyeb was executed, but Nasrullah isn't interested in hearing about that.

I ask Nasrullah whether he regrets the old life.

'What should I regret? We loved and drank and fought and stole.' His chuckle sticks in his throat. 'Then the mullahs came and the boot was on the other foot.'

❈   ❈   ❈

My alarm clock went off at a quarter to four. Bita didn't wake. I dressed and went down to the basement car park that we share with our landlord. Our Kia Pride was standing there. We'd bought it the week before, at a family discount from Bita's mother, with a free stereo.

I once briefly owned a Ninety-Eight Oldsmobile, a battleship. If you were in the back seat of the Olds, it was prudent to get your feet out of the way when the driver's seat slid back hydraulically. Mere shin wouldn't stop it.

Tap the Pride with your nail and its bodywork rings like an empty tin can. Changing gears is the clash of children's swords. The vanity mirror on the back of the passenger sunshade is an optional extra.

The Shah helped sink the Olds. The Ninety-Eight appeared in 1971. Two years later, the oil price hike turned big cars into an extravagance. Car-buyers asked a new question: 'How much does it do to the gallon?' If you were driving a Ninety-Eight, the answer was not much. Put your foot to the floor, and the swoon of the fuel gauge was visible to the naked eye.

Our Pride's relative old age, five, was its selling point. New Prides were more than 50 per cent Iranian-made. Five-year-old Prides, on the other hand, were mostly Korean. Apart from the steering wheel, which was off-kilter, and the handbrake, which didn't work on hills, and the stereo, which couldn't be turned down, our Pride did what it promised.

This distinguished it from the Daewoo Cielo that my mother-in-law had bought. The Cielo was shiny and sleek, with a metallic, olive-green finish. But it was considerably more than 50 per cent Iranian. It had required much 'fine-tuning' since lurching out of the showroom.

I reversed the Pride into the street, shut the garage door and joined Modarres, heading south. Modarres was full of rubbish that municipal employees were picking up. Beside a slip road, men were setting up a stall of stolen and salvaged car parts (Paykan gearboxes are a favourite). I picked up Mohsen, my assistant, in Martyrs' Square. We drove south, and Mohsen congratulated me on the car. He told me about the terrible time he was having with his Pride, which he'd bought brand new.

We went through Shush Square, where a lot of addicts score their opium and heroin, passed the Southern Terminal and entered a neighbourhood called Javadieh. It was five. After driving around for a while, we came across some people walking in the same direction. They were mostly families, or groups of young men.

Following the people, we rounded a corner. There was a floodlit piece of unevenly paved ground. To one side, about one thousand people stood watching a wall with barbed wire on the top. Neither Mohsen nor I knew what lay the other side of the wall, though we weren't far from the airport. Every so often, a light would go diagonally up into the sky.

We parked the car. A troop of riot police quick-marched, boots clumping. There were dozens of unshaven regular policemen. Some of them had hats with enormous peaks; they looked like a junta with a hangover. There were Intelligence people in rollnecks and leather jackets. They, and the floodlights, made the scene look totalitarian and hopeless.

A crane with a cab stood by the wall. Its chain had the kind of hook that you see on cranes around the streets, scooping up illegally parked cars and putting them on trailers to be taken to the pound.

In the crowd there were lots of young men. There were women in twos and threes. Lots of the boys were speaking Azeri Turkish; their families must originally have come from the Azeri provinces of northwestern Iran, and settled in Javadieh. The lads were joking and larking around. They wore a new wispy kind of sideburn.

Mohsen saw a man arrive with his small daughter and said, 'I don't think I'd bring my daughter to an event like this.' He approached the man, who told him that his daughter was mentally handicapped and wouldn't understand what was going on.

The crowd heaved. I was almost knocked off the mound of earth and bricks, ground that we'd occupied to be higher than the others. A car had arrived – an old Corolla with green police markings. The people behind me put their hands on my shoulders, trying to get a better view.

Rahmpour got out of the car. Everyone knew him. There were people who had been at school with him. Years ago, some of the women might have received a poem from him. He had curly hair and a chubby, dark face. Was that a smirk? He looked shifty.

Shifty enough to burgle his uncle's house late one night and, when challenged, to murder his aunt and two cousins, one of them a baby of forty-five days? (The answer probably depended on how badly Rahmpour needed the drugs. Very badly, I'd say.)

He was surrounded by officials and policemen. Mohsen identified one of them as the judge. Rahmpour walked back to the Corolla and sat down in the back seat. One of the boys in the crowd said, 'He's been pardoned.'

In a case like Rahmpour's, the victim's next of kin can pardon the killer and he gets a short prison sentence instead of death. (Thats in recognition of his debt to society.) That sometimes happens when the crowd thinks that the man has been unfairly sentenced and entreats the victim's family to be lenient. But that wasn't the case with Rahmpour. Everyone seemed to think that Rahmpour deserved to die. No one muttered, 'Poor Rahmpour!' or, 'It's a shame!'

It turned out that Rahmpour hadn't been pardoned. Later, in the evening paper, we read that he'd asked for a last cigarette. The paper published a picture of him sitting in the back of the Corolla, enjoying a last puff.

Behind us, a state television crew was interviewing people. Some politicians were uncomfortable with public executions. The television crew wanted to show that the people, by contrast, were in favour of them. Before the camera started rolling, the interviewer instructed the interviewee: 'When I give the sign, say you think Rahmpour should be executed in public and that you approve of public executions in general. Then thank the wise and esteemed judge for sentencing Rahmpour to public execution.' The man nodded.

Rahmpour got back out of the car. They brought him back to the crane. His knees looked strong. It *was* a smirk! A man put a red noose around his neck. 'That's his uncle,' one of the boys around us whispered. The uncle spat in Rahmpour's face.

The officials took a few steps back from Rahmpour. The crane's arm started to rise. We watched the slack tighten and Rahmpour rose into the air. The crane stopped when Rahmpour's feet were about five metres off the ground. He wriggled – first, his shoulders, and then his buttocks.

Silence. No coughing, whispering, scuffing of feet. No wind, plane, birdsong. The silence wasn't for Rahmpour, scum who deserved to die. But now he was hanging, he'd ceased to be a human being with particular characteristics. He became anonymous, a body. And his soul rose, or went sideways, or darkened the earth. It definitely went. I felt it. Everyone felt it. So the silence wasn't respect for the dead, but for death.

Mohsen told me they'd keep him there for a total of twenty minutes. Then, on the judge's order, he'd be brought down and examined. If he was still alive, he'd be hanged again, and, if necessary, a third time. (Three death sentences for three victims.) If, by some miracle, Rahmpour was alive after that, his survival would be taken as a sign from God. He'd be rushed to hospital and no effort would be spared to save him.

CHAPTER SEVEN

Gas

I don't remember more than a handful of old soldiers who, when I asked, said they hadn't been gassed. Some, like Mr Zarif and Alavi Tabar, had been lightly gassed. They complained of pains in the chest and shortness of breath. Amini had been gassed several times, but his facial scarring was diversionary and spectacular, and I almost forgot about the damage inside. It was only with the cleric Mr Rafi'i that I got an idea of the gas's patience; his pathetic cough suggested that Saddam's formulae were belatedly carrying him away. On my third trip to Isfahan, I met a military doctor. He estimated that the city had around eight thousand survivors of gas poisoning. Those who'd been very severely gassed had died soon after – from internal blistering or congestion of the lungs. In other cases, modern drugs and tracheotomies mitigated the effects, but not indefinitely. Now, fourteen years or more after their pain started, the eight thousand were starting to die.

In the summer of 2002, when I visited gas victims in Isfahan, American officials had begun citing Iraq's use of chemical weapons against Iran as justification for the attack they were planning against Saddam Hussein. For the victims, it was a predictable indignity. The Iraqis could never have gone chemical without the assistance of foreigners – and their moral selectivity. And yet, European chemicals companies professed ignorance about the uses to which Saddam intended to put the 'pesticide' plants they were building him. There was some queasiness in the Reagan administration, but this was subsumed by the strategic goal – to reopen diplomatic relations with Saddam and ensure that he didn't lose the War.

The Iraqis were using poisonous gas as early as 1982, but it wasn't until the following autumn, when the West German-built 'insecticide' factory at Samarra started making mustard gas, that production reached industrial levels. Later, this and a second plant, also made by West German companies, started producing deadly nerve agents. During Operation Kheibar, in 1984, helicopters that had been built by the Soviets, French and Germans dumped canisters of poison gas on the Iranians. Shocking reports circulated, but Iran didn't get the diplomatic support that other countries would have got. (You only had to turn on the TV to see the Islamic Republic inciting extremism or being accused of acts of terrorism.) In 1984, in the light of an investigation ordered by Perez de Cuellar, the UN secretary-general, Iraq was exposed as a violator of the 1925 Geneva Protocol outlawing the use of poisonous gas in wars. Member states looked the other way.

In 2002, I visited the Isfahan branch of the Foundation for the Dispossessed and War Disabled. An official explained to me that the war-wounded are examined and issued with a figure, a percentage, which rises in proportion to the gravity of their disabilities; the benefits they receive reflect this percentage. The official had never come across a percentage higher than seventy; that was for paraplegics and the victims of severe gas poisoning. While we talked, supplicants came and went. A healthy-looking man (10 per cent) pleaded with the official to intercede on his behalf; a state bank had refused to advance him a loan for the rebuilding of his house, which had been destroyed by fire. (The official said he would see what he bcould do.) He was followed by a middle-aged woman; her husband (60 per cent) wanted to know why the foreign car he had been promised as a palliative for his suffering hadn't turned up. The official replied that hundreds of veterans were asking the same question. The cars hadn't cleared customs.

The official took me to see the people at internal security. In the corridor, we came across a man who was staggering and shouting. The official asked me if I would like to visit an institution for mentally deranged veterans. I said I'd prefer to meet victims of the gas and their families. The official asked whether I had any preference, percentage-wise.

Internal security detailed someone to take me to see Mr Karimi. He was 70 per cent, and had the tremulous voice that I'd come to associate with people who'd been gassed. One of his two sons had a distended head. His wife wore a chador adorned with wild flowers and served us sherbet.

During an operation in 1982, he'd been ordered to watch over an abandoned Iraqi motorbike, to make sure it wasn't stripped for parts. Suddenly, he heard the sound of an Iraqi plane. It dropped its bombs fifty metres away. Instead of a loud bang and flying shrapnel, there was white smoke, in the midst of which were glowing shapes, the size of tennis balls – perhaps ten of them, drifting towards him. It was before masks, before awareness, but Mr Karimi ran. He tripped and fell over, involuntarily breathing in a gulp of gas as he got to his feet. Running made him sweaty, too; the gas entered his body under his armpits, and around his groin. There was a smell, like garlic.

By the time Mr Karimi got to the field hospital, his nausea and streaming eyes were giving way to intense pain, especially in his sides. He started coughing, which increased the pain. Blisters grew over his eyes, preventing him from opening them. He'd been poisoned by a blend of mustard gas and nerve agents. There was no antidote. Nurses applied pomades. Four days elapsed before he could open his eyes.

'You'll take tea,' said Mrs Karimi. She was carrying a tray with tea and *poolak*, discs of caramelized sugar that melt on the tongue. Mr Karimi didn't take tea. He rolled up his sleeve and injected himself with a drug that would allow him to breathe more easily.

Amid the poignant courtesies I received from the Karimis there was a current of recrimination. I was a Westerner, and they considered the West to be as guilty as Saddam for Mr Karimi's attenuated martyrdom. Mr Karimi showed me his pirated translation of Kenneth Timmerman's book, *The Death Lobby: How the West Armed Iraq*. I had the English-language original, which documents Western governments' dealings with Saddam before and during the War, their willingness to arm him, and the blind eye they turned to his production of chemical weapons. Like Timmerman, Mr Karimi found it hard to believe that the West German government had not known, as it claimed, that its companies were helping Saddam produce poisonous gas and nerve agents.

I observed that America seemed determined to topple Saddam. Mr Karimi had seen Donald Rumsfeld on television a few days before, demonizing Iraq. 'It was different in 1983,' he said. I looked puzzled, so he explained: 'That was when Rumsfeld went to Baghdad, and told Saddam that President Reagan wanted to expand military, technical and commercial ties.' Later on, back at my hotel, I looked up Rumsfeld in Timmerman's index. There was one entry; Mr Karimi had memorized it.

Betrayal came up a lot with Mr Karimi. At the time he said the betrayal had taken place, Iran was being accused of murdering and kidnapping Western military personnel and civilians. Iran had attacked Western ships in the Gulf and would attack many more. It was trying to export violent revolution across the Middle East. Mr Karimi couldn't sincerely think that the West had betrayed the Islamic Republic; there had never been a friendship to betray. Rather, I think, he was referring to an abstract betrayal – a hypocrisy, a betrayal of truth.

Mr Karimi had had the chance to go with other gas victims to Germany, for treatment at the Iranian government's expense. German doctors had developed a laser treatment, and they were good at tracheotomies. 'When it comes to chemicals,' said Mr Karimi, 'they're the best. They're the best at producing, and the best at treating.' He refused to go.

He worked as a taxi driver until 1995. Then, his lungs got worse; every few weeks, he would have to go to hospital and be put on a respirator. When he was at home, he read the papers and listened to the news, to see what would happen to Saddam.

He asked me if I wanted to ask his wife some questions. Mrs Karimi had been a young girl during the War. She and some school friends had vowed to marry men who had been badly wounded in the War. It was their way of doing their bit.

A lot of her energy, she said, was taken up refereeing between her husband and the boys. The boys could be boisterous and thoughtless. She had to make sure they did nothing to upset their father. 'We keep entertainments to a minimum.'

She didn't want pity. 'I accepted this responsibility, and the hardships are mine to bear.'

Sometimes, she said, Mr Karimi would get frustrated at being unable to do things. He would get upset and angry and take it out on the boys. Her husband nodded gravely. There were some unhappy homes, he said. Recently, a veteran had slit the throat of his two-year-old.

My last appointment in Isfahan was with a doctor at his surgery, quite late in the evening. He was an acquaintance of someone I'd met with Mr Rafi'i in the Rose Garden of the Martyrs. The first thing he told me was that he didn't want me to tape what he said. He'd prefer it if I didn't write his name and didn't take his photograph. Photographs and cassettes (audio and video): they're Iran's inert witnesses. At any time, enemies can bring them out and use them against you. The doctor didn't have suspicious eyes like Amini's – I felt, instead, that he was observing a cautious protocol. The idea that some foreigners might be interested in hearing about his experiences in the War surprised him. They'd never shown much interest before. So, the doctor would err – if it could be called erring – on the side of restraint.

He was handsome – not tall, but slim, with delicate features and a cleanliness beyond hygiene. (I'd been sensitized to such things; my enquiries into the War had brought me into contact with many people who were dirtied, and I longed for a spotless soul.) He had slender fingers and was soft-spoken. This combination presumably accounted for the abundance of female patients in the corridor outside. Feeling for lumps; peering some-where private; breaking bad news or good; if a man is to perform such intimacies on a woman – and Islam permits it, so long as there's no qualified woman to hand – he should be solicitous and pure of intent. These qualities shone from the doctor's little almond eyes and, in a strange way, from his brilliant white teeth. No man could have qualms about entrusting his ailing wife to such a physician. No woman could not but fall a little in love; the butterflies in her stomach would be settled by the virtue of his heart.

When I met the doctor, I was feeling sluggish and laid low – not only by the summer heat, but also by reading about the second half of the War. Casualty figures soaring into the realm of meaninglessness; the hurling without remorse of young men to deaths that were both banal and nauseating; turbaned invitations to martyrdom. For me, the War had become awful lists of battles, the battles in each list having the same name but a different number. After the name of the battle, there was a bracket containing another number; this denoted an estimate for Iranian casualties. Val-Fajr Eight, when Iran took Iraq's strategic Faw peninsula (more than twenty-five thousand); Karbala Three, an amphibious assault on Iraqi oil platforms, in which just 130 Iranians out of two thousand reached the objective. The account that Bita's family had told me of Tehran under missile attack seemed more immediate. A few hundred casualties; panic disproportionate to the danger; a blow to morale. It was more real than the abattoir of southern Iraq.

At the front, the feeling of lassitude had given way to frustration, and then to anger. I'd guessed it while talking to the old soldiers – from a sigh or an incautious criticism. It lay in their preference to talk about anything other than the second half of the War. Now, they had to devise means of getting through the aftermath. Amini had evenings with his friends, by the river. Mr Rafi'i had the soil between his fingers. Mr Karimi had his wife, and Timmerman. I wondered what the doctor had. He seemed the most together of all the veterans I'd spoken to. He looked and sounded healthy. Yet I'd been told that he'd been badly gassed. I asked him if that was true.

He'd been working in a field clinic, he said, set in a depression and roofed with corrugated iron, when someone shouted, 'Gas!' He'd forgotten his mask. He knew that gas, being denser than air, would sink into the clinic, but a man was blocking the entrance when he tried to flee. The man was badly injured; he needed first aid or he'd bleed to death. The doctor dressed the man's wound and gave him some drugs. The gas settled. (It was Tabun, a nerve agent, and it takes a minute or two for the nostrils to pick up its smell.) Doctor and patient started feeling its effects at the same time.

I asked him whether he regretted his decision to help the soldier. He said, 'No, God was helpful to me later on. If you'd come a few months ago, I'd have been unable to talk to you; my condition was much worse and I was coughing continuously. I had to go to hospital five or six times, for twenty days at a stretch. I was taking lots of antibiotics.'

Then, something extraordinary had happened. To the astonishment of

Isfahan's medical community, the doctor had made an almost complete recovery. Now, as long as he avoided strenuous exercise, he felt fine.

It was down, he said, to the 'grace of God'. Perhaps, I thought later, he felt that God had rewarded him for his selflessness in the field clinic. But there had been nothing preening about his description – no effort to justify or prepare me for the honour that he would go on to describe. Then, I thought of the worried women outside in the corridor, and his late working hours. Perhaps he viewed his recovery as an opportunity to do God's will – to heal and comfort others.

In view of the conditions he'd imposed on our conversation, he was surprisingly willing to talk about politics. Khomeini, he reckoned, should have dealt with Saddam less belligerently. He regarded the second half of the War as a tremendous waste. 'We should have accepted a ceasefire long before we did.' His readiness to criticize was partly, I think, a manifestation of the self-confidence that comes from being a professional in a country that greatly values professionals. But I sensed another, more personal, emancipation.

Some of the doctor's thinking was less imaginative, bound up with the tendency to blame foreigners for domestic catastrophes. He believed that Saddam had been a pawn; the Iran–Iraq War had been engineered by the West with the aim of bringing about the mutual destruction of the only Middle Eastern armies capable of standing up to Israel. The end of the War, he maintained, hadn't put an end to nefarious Western designs. Enemy governments supported the dealers who'd hooked millions of Iranians on Afghan opium and heroin. Through UN-sponsored family planning, they were trying to reduce the number of Iranians in the world. Although he spoke of Iran's current leaders without love, at least they were trying to prevent the country from falling back into America's clutches. 'If that happens,' he said, 'Iran will go back to how it was before.'

I asked the doctor how many children he had. He smiled: 'lots'. Then, he answered, 'ten'. They were from the same wife. Later in the conversation, something made me ask whether he'd gained anything from the War. I'd never asked that question before; in the company of most veterans, it would have been ridiculous. But my hunch was proved right. He answered: 'The fear left me. I became a man.'

While convalescing, he'd undergone some kind of religious conversion. That's not to say he hadn't been a devout Muslim before – rather, I think, his Islam had been semi-public and political, bound up with other people who prayed at the same time (watching the lips of the same Friday prayer sermoniser), with the same revolutionary intent. Then, towards the end of the War, the people had to make a choice; they could either follow the same lips as before, or seek their satisfaction elsewhere. Most people, frightened of insubordination, chose to follow the same lips as before.

The doctor turned to Islamic philosophers, especially to Alameh Tabatabai, the father of modern Shi'ite philosophy. Encouraged by what he read, he embarked on what he called a 'struggle against the ego' that brought him to a kind of Gnostic realization – a way of seeing God in everything, a way of forgetting his own petty fears. Gnosticism, or knowledge, he emphasized, is different from mysticism. (He mistrusted mysticism with its show-off dervishes.) His Islam was now self-sustaining, contained within a hard shell. Before, it had depended on external stimulae.

As I was leaving him, the doctor gave me two of Alameh Tabatabai's books. (I haven't started my struggle against the ego, so they sit unread on my shelf.) As I took the Tehran road out of Isfahan, I thought I noticed a chink in his Gnostic defences. 'Those who launch revolutions,' the doctor had told me, 'always feel the first blows.'

❈   ❈   ❈

'At that time', Alavi Tabar told me the next time I saw him, 'I was seriously political and a good speaker. I'd speak against the right and they'd call us the left because we supported nationalization of the economy and greater cultural freedoms, and they opposed us on both issues. They demoted me so I was just a normal Basiji. I said, "fine". I wasn't instinctively such a warrior anyway.

'Once, the Friday prayer leader came from Shiraz and addressed the Revolutionary Guard. He attacked a plan to distribute land to the poor and I spoke strongly against him, and in favour of the plan. There was a row, and the Revolutionary Guard booed him. The majority of the Revolutionary Guard was on my side, and that scared the commanders. I was out on a limb. They told me I was no use at the front. I did some training and was assigned to field intelligence.

'They sent me to a region just over the Iraqi border, in the dunes. We'd dug a trench on a rise, and we would send our reports from there. There was a road dividing our position and our forces. We could only cross that road at night; during the day the Iraqis would hit it. The Iraqis also had trenches among the dunes, and you never knew if you'd find Iraqis in them. Some nights, we'd go forward to gather intelligence, or we'd map the minefields and de-mine. There were six of us working in a team. We stayed in that sector for three months, and we went back for a bath once every two weeks.

'Once, I was returning from a night mission and I realized I'd left it too late; I wouldn't be able to cross the road before first light. I took cover in an Iraqi trench. As dawn broke, I saw that there were six dead Iraqis in it. My route back was blocked; there were Iraqis in trenches ahead of me. The nearest Iraqis were three hundred metres away. I was one, with a gun and sixty rounds. They were five. They had an RPG launcher and a machine gun.

'I buried the dead men, for reasons of religion as well as hygiene. One of them was really fat. Then, I set about repairing defences in the trench. I sorted out my food and rationed it. I vowed to myself that, in the last resort, I'd attack the Iraqis if there was an opportunity; I would have hated to surrender. That night I watched and listened, to determine if there was any movement from the Iraqis blocking my return. There was none. I slept.

'In the morning, I discovered that foxes had exhumed the bodies of the five Iraqis, and partly eaten them. I buried them again. I thought, "Who were they, and where did they come from? Where are their families now? What did they do to deserve this?" I tried to guess what their occupations had been. When I awoke on the third day, I saw the foxes had been at work again. The men had mostly been eaten, so I didn't bother burying them again.

'At dusk, I'd think about my wife. We'd struck a deal before I left, that we'd look towards the sinking sun and think of each other. I thought of my child. I thought a lot about what would come after death and about the last bullet hitting me – and what I'd see the next time I opened my eyes. I thought about the sins I'd committed but I was convinced I'd go to heaven.

'On my second night in the trench, I had a terrible dream. I dreamed they were interviewing my wife after my death. She was explaining all my faults. When I woke up, I found that I'd sweated buckets. I decided that if I was spared I'd change and do more to help people and that I'd go to deprived areas and abolish poverty and illiteracy. Sometimes I thought God had brought me to this trench to test me and then I thought: how puffed up you are with your own importance! Who are you that God should test you?

'On the third night, before dawn, there was the sound of gunfire. That could offer me cover, so I went. There was shooting going on between some Revolutionary Guardsmen and the Iraqis in one of the trenches that lay between me and safety. As I went, I hoped to God they hadn't mined the route back. I decided to pass close to the enemy trench; I couldn't be sure they hadn't mined the terrain on either side. But an Iraqi soldier saw me.

'We leaped on one another and we both dropped our guns and then he tackled me in the waist and rammed me into the ground. He was much bigger than me, but he was paralysed with fear. I boxed his ears with my hands and finished him off with my knife. The sound of gunfire had subsided. My hands were covered in blood. I sat there for a few minutes; it's not easy, killing someone that way.

'I started running. When I got close to our lines someone shouted, "Ali's alive!" They asked me whether I'd been hit and I said I didn't know and it turned out that I hadn't. I saw a friend and said I killed someone and he said there was nothing you could do. Life is important; it's worth staying alive. I was reminded of James Bond – you know, live and let die.'

❖　❖　❖

'A little later in the War, they didn't want me talking to the other lads, so I was put in supplies. I had a little store, and that was the lowest you could get. I would sell things to the lads. It was boiling hot; no one came to buy from me, so I became a bankrupt shopkeeper in the corner of the front.

'To start with, I'd take my fortnightly baths in the river. An Iraqi tank had toppled into it, trapping the water and creating a kind of pool. Then, one by one, the bodies inside the tank floated to the surface, and that put an end to my bathing in the river. After that, I bought a water container that could take two hundred litres. I jammed it in the trench and filled it with water. I would sit in it to escape the heat, and read. At dusk, I'd do my exercises.

'There was an old man with me, a Basiji. He'd come to the front as a volunteer. He had two wives and I'd say I wanted two wives, too, but what could I do to stop them killing each other? He had allergies to certain foodstuffs, and would make traditional potions for himself. At night, when he was sleeping, I'd prick him with a pin and he'd cry out, "A scorpion stung me! Scorpion!" Sometimes I'd write letters on his behalf, and read out letters that he received from his family. His family would write to him about their problems. He was a good man.

'At that time I was saying a lot of prayers. I was quite mystical, and the old man liked me because he knew I took particular pleasure from being on my own and praying, and he used to tell the others. I had a long beard and looked strange. Once he brought an old mullah and the mullah wanted to sleep alongside me. I spent much of the night in prayer. The next morning, the mullah kissed my hand and said, "Forgive me if there was a time when I had unfavourable thoughts about you."

'One day, the old man was hit by a bullet and his back was smashed up. It was several kilometres to the nearest first aid. I picked him up and ran with him. He was as light as a feather, but carrying him that far was quite an undertaking. After a while, I started vomiting. My kidneys were hurting like hell and I couldn't see. Later, he said, "You saved my life", and I replied, "Yes, I made a mistake; you were off to heaven and I got in your way." My back was sore for months after that.

'Every two weeks I'd go to the library in Ahwaz and borrow books. I read sociology and religious commentaries. I read all the Imam's books, and all Mottahari's and al-Sadr's and Alameh Tabatabai's. I read Popper with great interest, and I greatly appreciated Isaiah Berlin. I read American economists and French sociologists. British analytical philosophy helped me a lot; it taught me criticism and plain speaking. I had big problems with the Marxists; I couldn't understand what they were saying, while I felt as though I understood Popper perfectly. *Open Society and its Enemies* was a huge influence on me. It acquainted me with the great danger of totalitarianism, and totalitarianism became my subject.

'They wanted me away from the lads, in case I corrupted them. I understood democracy, which they didn't. They believed there should be popular participation in government, but no competition. We'd say that democracy means both participation and competition. That was a controversial point of view, and the commanders were scared of me. Some of them were frightened to be on their own with me. They'd say, "You're capable of bewitching us."

'When I was saying my prayers, no one would come and kneel next to me; someone might see them. When I was on leave in Shiraz, they'd say, "He's devout, but he says strange things." They'd ask me, "Do you think you understand things better than the mullahs?" If I got into an argument with a mullah, I'd ask him which books he'd read on a certain subject, and I'd invariably have read the books he'd read on the subject, and others besides. When it was clear that I knew more than him, I'd say, "Well, you ought to be emulating me!"'

It was early one morning in the summer of 2003. Mr Zarif and I were in Abadan, the British-built refinery town that lies adjacent to Khorramshahr. We were standing on the deck of a fishing boat that was moored to the bank of the Arab River, and Mr Zarif was using his new acquisition, a Russian Zenit. He took several photographs of the wheelhouse with its home improvements – a scarlet carpet over the hatch that opened onto the engine; plastic flowers in a vase that was secured with twine; a very old cassette player for playing vulgar tunes. He captured wiry Arab fishermen standing on deck while ice blocks the size of a child's coffin slid from a truck on the quay down into the hold. He surveyed the ragged fleet of boats with their Iranian flags straining high on the mast. Then, framing the Iraqi shore in his viewfinder, he had one of his turns. I had to ask the driver, Rostam, to take us back to the hotel so Mr Zarif could have a lie-down.

On the flight from Tehran, we'd talked about the Abadanis. Before the

Revolution, I had heard, they were reputed for their hedonism. Mr Zarif whistled. And how! When he was a child, the Zarifs had been on a family holiday to Abadan. Strolling in the evening, they'd seen groups of Abadanis, men and women together, dancing around a man with a pair of drums and a good voice. There had been Gulf Arabs, drawn across the water by casinos and nightclubs that were illegal in their own countries. On a Saturday night, Iraqis from Basra would gather in cafés on the far shore of the Arab River – where they could pick the sexy films that were broadcast on Abadan TV.

The town had a strange, threatening variety. Apart from the Westerners at the refinery, you found Armenians and Jews and even Indians. (The Indians were leftovers from the British era; they had a Sikh temple!) When the War started and Abadan was levelled, people observed that God was punishing the Abadanis for being Westernized and debauched. Even now, they tell of the Abadani who couldn't care less that his aunt is loose and his father a thief, but hits the roof when he finds out that his Ray-Bans are fake.

Now, as then, the government's policy in Abadan is to promote Persians over Arabs, and to cut deals with the Arab tribal chiefs who wield influence over the hinterland. Now, as then, the Persians consider themselves a cut above the Arabs, deserving of the British bungalows and lawns that are the privilege of the bureaucratic and refining elite.

Rostam was such a Persian. He was a rich kid; his family lived in a bungalow and he drove his father's Pride saloon (not really a saloon – more a distended hatchback) for pocket money. He had no plans to go to university. He was waiting for his British visa to come through; he'd married his first cousin, a British citizen, and would get a job as soon as he reached England. (You'll be waiting a long time, I felt like telling him; British immigration is wise to the fabricated marriage caper.) When I asked if he knew Arabic, he replied, 'I understand but I don't like to speak it. I'm more interested in English.'

We went out again when Mr Zarif had got over his turn. Driving through the city, Rostam pointed out a house that until recently had an Iraqi tank in the driveway. 'Every so often,' he said, 'the guy who kept it would start it up and give it a good rev, just to keep the engine ticking over. Last year, they got wise when he tried to sell the engine to a barge-owner. They came and took the tank off him.' A bit further on, we saw a roundabout that had a statue in the middle, of an old man on a bicycle. During the War, Rostam told us, the man had rushed on his bike to warn the authorities about an unexpected Iraqi advance, saving the city.

We emerged onto the Ahwaz road with its sandy flats that flood in the spring. Rostam turned up the air conditioning and put on a cassette by Darioush, an Iranian pop star based in LA. (That was another difference; if we had an Arab driver, we'd have got a cassette from Beirut or Cairo.) Mr Zarif said, 'The last time I came along this road, I was holding a Kalashnikov. Now, I'm holding a Zenit.' He took a photograph of one of the road signs. 'We used to shoot them up as we passed, for target practice.'

I said: 'What else do you remember?'

'I remember the look of resentment on the faces of the local Arabs when they saw us – as if we were foreigners.' The Isfahanis had been fired up. The Arabs had been unenthusiastic about having a war fought in their corner, over resources that they considered their own.

Rostam looked at us in the rear-view mirror. Our conversation had confirmed what he had probably suspected: that Mr Zarif was a veteran who had come to visit wartime haunts. In a mocking tone, he said, 'Khuzistan has become a pilgrimage place.' Mr Zarif replied, 'In that case, everywhere should be a pilgrimage place, because there's been a war everywhere.'

After driving for an hour, we came to a small, desolate place on the banks of the River Karun: Darkhoein. As we drove through the village, Mr Zarif complained about how he couldn't get his bearings or remember where anything had been. We asked two or three people, but they hadn't heard of the Division of the Imam Hossein. We got out of the car and stood for a few minutes on the river bank, near the point where Iran's forces had made the nocturnal crossing that marked the beginning of the assault on Khorramshahr. People were crossing on a barge, attached to a tensile cord that spanned the river, propelled by a straining outboard motor. Mr Zarif was trying to remember whether, after Khorramshahr, Hossein Kharrazi had moved his HQ across the Karun. He asked the people standing alongside us. They looked at him blankly. He muttered, 'Don't tell me it was all a dream.'

Eventually, an old man directed us to the atomic complex that the Iraqis had occupied at the beginning of the War – it had been Hossein Kharrazi's first HQ. There was a heroic hoarding at the gate. It depicted a smiling Hossein next to a Basiji who was kneeling in prayer. There was a quote from the Imam: 'Blessed are those who are seated on the wings of angels.' A little further on there was an Iraqi tank, a beetle-like T50, rusting on a plinth.

After that, we went up and down the Abadan–Ahwaz road, trying to find a village called Muhammadiyeh, where Mr Zarif had learned to lay television mines. In a village that may or may not have been Muhammadiyeh, we disturbed a woman in a hovel; she told us her husband would be back later and slammed the door. Further on, there was a detachment of Basijis in a sun-blasted caravan. They weren't volunteers, as Mr Zarif had been, but conscripts. They crowded out of the caravan when they heard our Pride, tucking their shirts into their trousers and rubbing their eyes. If their superiors caught them sleeping, Mr Zarif said, their military service would be extended by at least a month.

Back in Abadan, Mr Zarif wanted to take photographs of the old British residential area. There was an enormous cinema from the 1930s; seeing it gave me an odd and unexpected pang of homesickness because it was very similar to a cinema I'd gone to in London as a child. Some of the houses were flat bungalows of the kind you find in an Indian cantonment, but there were also fantasy cottages with wobbly marzipan transepts and tiny windows hidden by peppermint shutters under tipping roofs. Mr Zarif was very interested in one cottage, adjacent to a dyke overlooking the river, which had been half bombed and not repaired. While he took close-ups of turquoise tiles on brick kiosks overlooking a suburban green, I tried to explain the importance of the box hedge to British life.

As we were driving back to the hotel, he said: 'I don't know if I didn't see the British neighbourhood when I was last here, or if I saw it and it didn't mean anything to me. But I have no recollection of these buildings.'

That was 1988, when the War and jurisprudence obscured everything. Now, he was turning into an artist. He was nearing the end of a scriptwriting course. He was making notes for a novel. He was becoming a photographer. Amid the pleasure of new discoveries, some of them within himself, there was a wistfulness and regret. The doctors had botched an operation to remove a tumour from his mother's brain and she was fading.

He'd taken rolls and rolls of the Imam Hossein ceremonies. Then Bita had introduced him to a photographer friend, Mehran Mohajer, who advised him to get down among the details. A few weeks later, Mr Zarif had shown me some photographs of a beautiful Qoran. There was a perpendicular view of the book lying open; the edge of each page was shown in fantastic detail, following the contours of the one below – and so on, like the rings of a tree trunk. There was another close-up, this one from an oblique angle, of the words themselves, and the sinuous quality of the Arabic turned them into features on the rolling landscape of the page.

Crossing the lobby, we stopped for a few moments to look at a model of the hotel as it had been during the siege of Abadan. (Despite being next to Khorramshahr, the city hadn't fallen, which turned it into a symbol of resistance – Iran's Stalingrad, Khomeini liked to say.) There were anti-aircraft emplacements and the craters left by artillery shells. An ambulance stood where Mr Zarif and I had sat the previous night smoking a hookah.

Waiting for our food in the empty restaurant, I asked him if he'd been disappointed by our tour that morning. A lot of the people we'd met had seemed indifferent to the War, or ignorant of it. Should they have been more appreciative?

He shook his head. 'The War has become muddied, Mr de Bellaigue. It's no longer clear in people's minds. The War was bound up with an ideology that an increasing number of people don't like.'

I thought of Rostam. He hadn't been born by the time Khorramshahr was liberated. He probably drew his first breath in one of the distant cities to which middle-class Abadanis had fled during the War. I doubted if he believed that the War had been fought in his name. Now he dreamed of making a life in England – there, the War would mean even less.

Mr Zarif said: 'They know other things now in Darkhoein – things other than the Division of the Imam Hossein. They know where the water is sweet and the location of the electricity substation. They know the places that are infested with snakes.' He was saying, I think, that human beings adapt according to what seems important now, throwing away old information.

When had Mr Zarif understood that the ideology lay defeated? Was there a moment of lucid agony? Perhaps not one moment, but a million moments spread over ten dull years. Now the realization had passed into him, replacing the ideology at his core. The way he mixed the butter and bitter brown *soumagh* powder with his rice, clicking his spoon and fork together, reminded me of the equanimity he'd shown in Qom: an ability to distance himself from him, before.

The kebabs were filthy, recently defrosted and served with insipid rice cooked with yesterday's oil. There was a basket of raw herbs as an accompaniment.

Mr Zarif, I knew, believed it was the fault of the big shots; he and the other young lads should have been better led. But that sounded like a negation of responsibility. He had terrorized teachers, stifled opinions, bullied. Was that someone else's fault?

I said irritably: 'We're on the coast and they give us frozen veal.'

He said genially, 'At least it's not European food.'

I frowned. 'What do you mean?'

He was still smiling. 'You said there was nothing so bad as European food.'

'I never said that.'

'Yes you did. You said that European food is tasteless and overcooked.'

'I never said that,' I snapped. 'I said that British food is tasteless and overcooked.' I went on recklessly. 'Much of the rest of Europe has a food culture that's a thousand times more sophisticated than Iran's.'

Mr Zarif couldn't believe what he was hearing. He said, 'Iranian food is among the best in the world.'

'How do you know?'

'I know the Arabs' food very well. I've been to Syria and Saudi Arabia and Iraq. Here and there, I've eaten lots of Western food. We have pasta here, you know.'

'That stuff! You call that pasta? You'll be telling me next that Iranian pizza parlours actually produce pizzas!'

I seized a basil leaf from the basket. 'Look at this! In Iran, you just pop this into your mouth and eat it. But have you thought what it would taste like if you crushed it in a pestle and added olive oil and butter and mature cheese and pine kernels and mixed it with home-made pasta? Can you imagine how delicious that would be?'

Mr Zarif didn't look at all sure that what I'd described would taste well. But he conceded that he wasn't informed enough to compare Iranian food with the food you found in the rest of the world. I felt better, but still resentful at the way the Iranians thought they knew best. I pushed my plate away.

A few minutes later, Mr Zarif finished his kebab. I said, 'You have a child, Mr Zarif, and he's growing up fast. What kind of Iran would you like him to inhabit?'

He'd disregarded our contretemps. 'To answer that question, we need to understand what kind of Iran exists now. As a nation, we have a problem that's reflected at the highest levels. But I don't think it's the state's fault. It's a reflection of the people and their contradictions.

'When I get into my Paykan and it lurches and coughs, I think to myself that the men who made it aren't well enough trained or paid, and that they have bad equipment and are badly managed and didn't sleep well last night. Perhaps they have problems with their wives. On the few occasions that I've been in a Mercedes and been astonished by its mechanical perfection, don't you think I've asked myself if the men who built this car are better off?'

I'd had similar feelings. Why doesn't anything work? Why does nothing happen on time? Why is everything so crappy and falling apart? Is it useful to spend so much energy mourning a man who died more than thirteen hundred years ago?

But there was a distance between myself and Iran. I would never be Iranian. (You cannot become Iranian – not spiritually. You have to be born one, like a Hindu.) But Mr Zarif was Iranian, and so these thoughts were acute, a kind of self-flagellation.

'Mr de Bellaigue, I was brought up to believe that a Muslim who doesn't say his prayers has no value. Now, I ask myself: it is possible that people love God as much as I, and yet don't say their prayers?'

A man who doesn't say his prayers does me a kindness. Is it less of a kindness than the kindness done me by a man who says his prayers, or the same kindness? (Or, conceivably, a greater kindness – some men who say their prayers do kindnesses only because they hope it will stand them in good stead when they stand at the doors to heaven.)

'Perhaps Mr de Bellaigue has the same appreciation of poetry as I, but doesn't say his prayers. I have friends who say their prayers with perfect regularity but don't appreciate the films of Mohsen Makhmalbof or the photographs of Mehran Mohajer.'

He'd mentioned my name casually, as if to suggest that the name wasn't important – it was an example. But it was deliberate, I knew it. He was telling me that his friendship with me hadn't been unproblematic.

'My environment now seems full of colour, no longer black and white.'

'Is that a nice feeling?'

'It's surprising.'

'And your wife?'

'She's surprised, too.'

It was bound up with the challenge facing his generation: to solve the contradiction between modernity and tradition. 'When I speak to my son, I want us to understand each other.' He was proud of Ali. He didn't want to stop being pals.

'Look, Mr de Bellaigue – I'm Iranian. I can't remember drinking alcohol or looking lustfully at any woman other than my wife. I can't remember going without saying my prayers. Behind this lies a thought, an essence, and this essence has to be made to harmonize with modernity. Then, our problems will be solved.'

Our tea came. A dead bag in tepid water. Mr Zarif said, 'What is the effect on the world if I don't say my prayers? Have I affected the balance?'

In the West they say that freedom is thinking and doing what you want as long as it doesn't harm or upset others in a way that can be crudely measured by a detective, medical doctor or psychologist. To Mr Zarif, this is absurd. Is it reasonable that a thought can remain inside, hermetically sealed? No! Is it possible to measure the harm done to society by an unhealthy thought? Of course not! This much is certain: the thought seeps into your actions, into the atmosphere. How ridiculous to imagine that a thought leaves no trace.

Early the next morning, Rostam drove us through Khorramshahr with its revolutionary daubs and murals. During the recent war between America and Iraq, he said, groups of Arabs had come into the streets and shouted pro-Saddam slogans. We went past a message written on a wall: 'Iran's strength lies in its martyrs.' Then we headed out of Khorramshahr, towards the border crossing at Shalamcheh.

We passed some warehouses that had been in shreds since the War, and decapitated date palms. A little further on, there was a column of trucks by the side of the road. 'Iraqi refugees,' said Rostam, 'waiting to go home.' The refugees squatted in the shade offered by the trucks, which were packed with their personal belongings. In a few hours they'd be in Basra, standing outside homes that had been inhabited by other people for a decade at least.

A young commando came out of the guardhouse and told us that the road was closed to anyone without a pass. I whispered to Mr Zarif: 'Tell them you're a veteran.' He hissed: 'If I said that, they'd be even less likely to let us through. Did you forget what I told you yesterday?' The commando said that an officer would be along shortly to authorize the Iraqis' passage. If we were in luck, he said, he might let us go on as far as the actual border.

We got out of the car. A kilometre or two to our left we saw ships plying the Persian Gulf, crowding the mouth to the Arab River. I looked ahead, towards Shalamcheh. Somewhere along this road, I thought, Hossein Kharrazi supervised the removal of a burning tank.

What if he'd survived?

At the end of 1986, Rafsanjani gave a bellicose speech to one hundred thousand Revolutionary Guards and Basijis. An offensive was in the offing: Karbala Four. The plan was to seize some islands in the Arab River, land on Iraqi territory and advance north along the Basra road. Thousands of newly recruited Basijis would take part, accompanied by seasoned soldiers from forces like the Division of the Imam Hossein. Kharrazi had trained and equipped an amphibious unit. The offensive would be launched with a surprise nocturnal attack by his frogmen.

Kharrazi's intelligence people had told him that the islands were covered with machine-gun nests. The far bank was a network of barbed wire and minefields. Iraq's tank brigades would be swiftly brought to bear on any attackers. Iran's artillery wouldn't be able to provide more than a limited barrage from the opposite bank.

Kharrazi made known his apprehensions, but the top brass wasn't listening. Kharrazi had stopped being the golden boy. He'd visited Ayatollah Montazeri to complain about the conduct of the War. His refusal to waste men's lives had pitted him against people higher than him. Rather than keep them behind the front, where they would be exposed to enemy shells and bombs, Kharrazi was holding most of his men in Isfahan, bringing them forward by helicopter when an offensive was imminent. There was a widening cultural gap, too. The government had started providing senior commanders with cheap loans so they could go into business. Kharrazi was uneasy at the thought of holy warriors getting rich.

Rafsanjani came to Ahwaz wearing a military uniform and invoked his authority as war supremo. Karbala Four would go ahead as planned. A few hours before the operation, there was a three-way radio exchange between the Revolutionary Guard HQ, Kharrazi at his field HQ behind the front line and his top commander. Kharrazi reiterated his opposition to the plan. HQ replied that the attack must go ahead as planned. Kharrazi conveyed the order to his commander. 'Today is Ashura,' he said, his voice cracking, 'and this is Karbala.'

And so it was. When Kharrazi's frogmen landed on the islands, it wasn't a surprise; they were lit up by searchlights and mown down. The following morning, the main body of Iranian troops crossed the river. Wave after wave advanced against fully prepared Iraqi positions. Within forty-eight hours, the Iranians had been hurled back across the Arab River, losing at least nine thousand dead. In Tehran, victory was duly proclaimed.

The officer we were waiting for arrived shortly after nine o'clock. It looked as though we were in luck. One of the commandos had to relieve a colleague who was manning a second checkpoint, a short distance up the road towards the border point at Shalamcheh. It was agreed that we would drive the soldier as far as this second checkpoint.

On the way, Mr Zarif pointed out primitive field telephone wires along the side of the road. (You couldn't eavesdrop on them as you could a short-wave radio.) There was a sign indicating that the terrain on the right had been de-mined, but Mr Zarif was not so sure. Over the several years that had elapsed since the minefield was mapped, he said, flooding would have moved the mines around.

A bit further on, the commando said to Mr Zarif, 'You were here in the War?' Mr Zarif smiled: 'Yes, in the Division of the Imam Hossein.' The commando said: 'That Isfahani . . . Kharrazi. His son was here just a few months ago.'

That would be Mehdi. He must be sixteen now. I'd seen a photograph of him in his grandparents' house, when he was very young, being kissed by a man who went on to become the foreign minister. Hossein and his wife had got engaged in the teeth of opposition from Hossein's mother, but Mrs Kharrazi had come round to her daughter-in-law. After Hossein's martyrdom, the girl had immediately contracted a second marriage, to Hossein's younger brother.

After Karbala Four, Hossein returned to Isfahan and it was clear to many people that he'd lost his will to live. He had no outstanding religious obligation to perform. He had been on the pilgrimage to Mecca. The Revolutionary Guard hadn't forgotten his opposition to Karbala Four. When it came for him to go back to the front, the authorities didn't put a car at his disposal. He went by bus.

We'd reached the second checkpoint. The commando thanked us for the lift. I said, 'But we haven't reached the spot yet!' Mr Zarif shook his head. 'No, it must be a bit further on.' We were only minutes from the border and the commando wouldn't let us drive on. He could get into trouble. Rostam turned the car around. I said, 'We may as well go to the airport.'

After Hossein Kharrazi's martyrdom, the authorities in Isfahan declared three days of official mourning. It was a splendid affair; the prime minister was on hand, and plenty of other bigwigs. It seemed as though the authorities had forgotten that, just a few days before, they'd sent Kharrazi back to the front by bus. But the mourners hadn't. Everyone knew lads who had been needlessly martyred in Karbala Four. Everyone had heard about Hossein's radio conversation before Karbala Four. After the funeral, lots of the old Basijis went to the front. It wasn't that they'd been galvanized into wanting to fight. There seemed to be nothing else to do.

At the beginning of 1988, Iraq's superior equipment and morale allowed

Saddam to launch some major offensives. Iran tamely gave back Iraqi territory it had spent months and thousands of lives to win. Rocket attacks on cities were causing panic. Morale dipped further when an American cruiser in the Persian Gulf downed an Iranian civilian airliner with almost three hundred people on board. The Americans said it was a mistake. Khomeini said that Iran must be prepared for a 'real war'. No one was.

Everyone remembers where he or she was on the summer day in 1988 when Iran agreed to implement a UN ceasefire plan that had been on the table for a year. (There were heart attacks in Tehran's bazaar; the hoarders had lost millions.) Three days later, Khomeini said, 'Had it not been in the interests of Islam and Muslims, I would never have accepted this, and would have preferred death and martyrdom instead. But we have no choice and we should give in to what God wants us to do . . . I reiterate that the acceptance of this issue is bitterer than poison for me, but I drink this poisoned chalice for the Almighty and for his satisfaction.'

Shortly after Iran's announcement, Mr Zarif went back to the front. Saddam was posturing; there was a fear that he would try and grab some territory for barter during negotiations. In the event, Saddam contented himself with sending thousands of deluded Mujahedin militants, who had taken refuge on his territory, on a suicidal mission deep into Iran, where they were massacred.

Mr Zarif was at the front when Saddam eventually accepted the ceasefire. 'From the Iraqi side, there was whooping and firing into the air. They were dancing. On our side, everyone was crying.'

❁　❁　❁

It was almost six. The temperature had cooled and people had entered the Rose Garden of the Martyrs – to clean graves, lay flowers and pay someone to sing an episode from the martyrdom of the Imam Hossein over their dead sons. Women lowed, while children ran around playing, and men and boys distributed cold drinks as an act of pious charity.

Amini was standing at the entrance. It was a year since we'd sat on the river bank with Zahmatkesh and the others. He'd been surprised to hear my voice. I put my hand up but he didn't see me. He was standing still and his lips were moving; he was greeting the martyrs. He entered a baldachin that contained the tomb of an important mullah.

I sat down on a bench. I'd bought a Tehran newspaper that morning but hadn't read it. I took it out of my bag. The front page bore a photograph of Ayatollah Taheri.

He'd coordinated the city's War effort. His son had been martyred while fighting in the Division of the Imam Hossein. He'd been an ally of Ayatollah Montazeri. Then, in 1989, Montazeri had got in trouble for criticizing the regime – for criticizing the execution of thousands of Mujahedin prisoners, and for the denial of people's rights. Shortly before his death, the Imam had stripped him of the succession. Taheri had been isolated, too; he was an ally of the disgraced ayatollah.

Taheri had been mostly discreet in his support for Montazeri. Now, however, he had issued an open letter explaining why he was resigning as Isfahan's Friday prayer leader. The people, he said, had shed 'pure blood' to 'establish the just government of Ali', but things had turned out differently. The regime depicted by Taheri was corrupt, opportunistic and thuggish. It was one of the most sensational critiques a member of the regime had made since the Revolution.

I looked up; Amini was by my side. We shook hands and he sat down. I showed him the newspaper. 'What do you think?' He shrugged.

I'd hoped to talk to him about several operations in which the Division of Imam Hossein had taken part. I'd hoped to learn how he felt about the changes that had taken place in Iran since the end of the War. But Amini had decided not to talk; he still didn't trust me.

Most of the veterans I had met in Isfahan would agree with Taheri; they knew that today's Iran was a parody of the Iran that Khomeini had promised them. The country had never known such moral corruption. Pre-marital sex, divorce, drug addiction and prostitution had reached levels that you'd associate with a degenerate Western country.

Amini looked at the photograph of Ayatollah Taheri and I experienced a recrudescence of the old frustration. He wanted to talk but his distrust of foreigners prevented him from doing so. He had only come to be polite.

I said: 'I recently read a novel that's set in London after World War I', and went on to describe the guilt and distress felt by Septimus Warren Smith on his return from the front.

Mr Amini said: 'What happens to him?'

'He jumps out of a window.'

'Islam doesn't countenance suicide,' he said with satisfaction.

I said: 'Did you take part in Karbala Four?'

'Why do you ask questions like that?'

At length, Amini said, 'Let's say hello to Hossein.'

We walked through the Rose Garden of the Martyrs. He started to talk.

'A few weeks after Karbala Four, Hossein's in a trench up at Shalamcheh and he's hungry and he says: "Boys, what have we got to eat?" And one of the boys says: "We've got chicken from last night; I'll heat it up for you on the gas stove." As it's warming up, Hossein sticks his head out of the trench and sees a water tanker take a wrong turn and head off towards the enemy lines. Any other commander would have sent someone to stop the water tanker but Hossein doesn't think; he gets up himself and runs off towards the tanker, shouting: "Stop!" The tanker stops and the driver sees it's Hossein, and he's thrilled. He gets out to kiss his cheeks and they're standing there and a shell lands and cuts Hossein in pieces.

'The boys rush out of the trench and see Hossein in pieces and they send him to the clinic, but it's too late; Hossein's gone. They go back to the trench and Hossein's chicken's still on the stove, cooking. It starts burning and everyone's sitting there. So, they sit there while Hossein's chicken burns, and the walls of the trench go black with the smoke, and they carry on sitting there until the stove runs out of gas.'

Mr Amini was sucking in, making a rasping sound. It's the sound badly gassed men make when they're trying not to cry.

We'd reached Hossein Kharrazi's grave. Amini leaned down, touched the top of the grave, and said a prayer in Arabic.

CHAPTER EIGHT

Parastu

It so happened that the sun had a chariot keeper, an old man with a white beard that was long enough to get caught up in his arms and feet, and that, being cotton-like, was useful for polishing the chariot. One day, the sun gave the chariot keeper his instructions from the Master. 'The Master wants you to clear out the rubbish that's accumulated in the celestial closet; he wants you to burn it or throw it away. Most important,' he said, 'you're to take out all the stars that belong to his subjects on earth, and send them down there. The Master wants everyone to have his or her own star.'

The chariot keeper grumbled. 'You think that's going to be easy? They've been chucking things in there for at least five hundred thousand years.' He couldn't see why the people on earth – to whom the Master had given a flattering title, 'the honourable species of humanity' – should be rewarded with stars in the first place. 'As far as I know, the honourable species of humanity understand nothing except killing and oppressing each other.' But he did as he was told.

The chariot keeper threw out tablets that prophesied the future, and the broken old wings of cherubs and angels. He came across burned-out stars and lightning bolts that had never reached their destinations. There were piles of old manuals associated with different gods, and (in an adjacent room) icons to represent them: tree gods, snake gods, sun gods, moon gods and gods that were human (with and without wings). The chariot keeper did away with Ishtar, Isis, Apollo and Venus. (He took care, however, to retain Venus's pitcher of holy water.) He smashed the god-king Gilgam-esh to smithereens and blew him away with a puff. Then, he gathered up documents that related to holy places and mountains, and to prescriptions that the Master had prepared for the people on earth over the past five hundred thousand years. He took them to a corner of the sky and clapped his hands to produce a spark, and they went up in flames.

Every morning, the chariot keeper would sweep up the stars and put them in a cupboard. You can't be too careful with stars; it won't do to have the sun or some angel turn up and start playing marbles with them. One evening, standing by the cupboard, he shouted: 'Hey! Come and help!' At first, there seemed to be no response. Then, millions of cherubs started to emerge from the four corners of the sky – they held sacks, and ladders made from the sun's rays. Under the chariot keeper's supervision, they filled the sacks with stars from the cupboard; each cherub was given the addresses of the people who were to receive the stars in his sack. The chariot keeper sealed the sacks and the cherubs set off, flinging their ladders down to the earth and descending with their sacks full of stars. As he watched, the chariot keeper thought it was the most beautiful sight he had ever seen – even more beautiful than the day the Master had ordered that every water lily in every pond should flower.

For a while, the chariot keeper fretted that the Master would have no stars on his cape. But the sun reminded the chariot keeper that this wasn't his concern, so he thought of ways to occupy himself while he awaited the cherubs' return. First, he combed out his enormous beard so it covered the whole sky. Then, he broke Venus's pitcher of water and scrubbed himself from head to toe. That made him feel young, and the river in the heavens echoed with his laughter. (On earth, the sky became cloudy, and then there was torrential rain and the sound of thunder.)

On the fourth day, the tops of the cherubs' ladders, followed by the tops of their heads, appeared in the heavens. When all were assembled, the cherubs confirmed that, as they had been instructed, they'd presented a star to everyone on earth, uttering the words: 'I entrust this star to you. From now on, know that you are free.' The chariot keeper asked how the people had reacted.

Different people had reacted in different ways. Children, for instance, had been delighted. Their eyes had sparkled and they'd started playing with their stars. The old people, on the other hand, had complained at receiving their stars too late in their lives. Most of the young and middle-aged people had been mystified. 'However much we explained,' one of the cherubs said, 'they didn't understand what the Master was getting at. Some of them lost their stars very quickly. Others hung them around their necks.'

Only a small group of people had got the point. They told the cherubs that the stars were unnecessary. 'We were free from the beginning,' they said. Such people had a complicated way of speaking – few of their fellow humans, let alone the cherubs, understood much of what they were saying. They were working on dictionaries, getting rid of words like 'predestination' and 'fortune' and 'destiny' and 'statutes', and creating new words, whose roots were 'free' and 'freedom'.

The chariot keeper said, 'I think one day I'll go down to earth and see what's going on. It sounds as though it's worth a look.'

❂   ❂   ❂

I read the tale of the chariot keeper and the cherubs, which I have severely abridged in the retelling, in a Persian novel by a woman called Simin Daneshvar. The novel's title, *Soo va Shoon*, is difficult to translate; it refers to a mourning ceremony for Siyavash, a hero of Firdowsi's The Book of Kings. Bita had recommended *Soo va Shoon* to me; she'd first read it a few years after it came out in the early 1970s. (It may have been, she said, among the books that she picked out of her grandfather's library in Isfahan and read at the far end of the garden during the summers she spent there.) In *Soo va Shoon*, the tale of the chariot keeper and cherubs is composed for the twin daughters of Zari, the novel's heroine; one of them becomes tearful if there aren't stars in the night sky.

The book is a description of female resistance and duty; Zari's capacity for love, and her mental strength, place her among the pre-eminent heroines of Iranian literature. For much of the narrative, Zari is engaged in an ominous struggle with her husband, Yusuf, over the extent to which their family – its coherence and security – can justifiably be endangered to further a cause. Despite her own political consciousness, and the activism of her husband, Zari tries to bolster the family against tremors from outside. 'They can do what they like,' she says, referring to groups that are trying to control the couple's hometown, Shiraz, 'as long as they don't bring their war into my nest.' To Yusuf's unhappiness at the city's deathly and deserted aspect, she responds: 'My city, my country, is this house alone.' But her indifference turns out to be unsustainable.

The Iran depicted in *Soo va Shoon*, which is set shortly after the country's occupation by Britain and Russia and the abdication of Reza Shah in favour of his son, has rarely seemed so fragile, so undeserving of nationhood. Back in World War I, local landowners and white-collar opportunists cooperated with the Germans who briefly controlled southern Persia at the expense of the British. Now, the same men are polishing British boots, selling grain to feed alien ranks (Indians, mostly) while some of their compatriots go hungry. As much as the perfidy and spinelessness, it's the myopia behind these expediencies that upsets Yusuf. During a society wedding that becomes a summit of fawning Anglophilia, he asks his wife: 'How come calves kiss the hands of men who are about to slit their throats?'

Soo va Shoon is a book of humiliations, some of them small and graded according to the hierarchies of local society, others larger, more shocking. These humiliations remind Yusuf and Zari that, despite being landowners with the additional advantage of a modern education, they are in thrall to foreigners, and to Persians who have gained influence and wealth by selling out to foreigners. That's not to say that there's an atavistic dislike of foreign people or cultures; Yusuf has studied 'abroad' – which probably means Europe – and Zari's father was reputedly the best English teacher in Shiraz. But, at the wedding party, we feel for them as they view a five-storey wedding cake that is surmounted by the Union Jack. We feel for them when they must put on a display of esteem for a fat foreigner named Zinger.

Zinger's name intentionally recalls the name of the sewing-machine company, Singer, for which he used to work as a salesman on commission ('no trousseau complete without one'). Now, preening in an officer's uniform with pips that Zari suspects are self-awarded; he mingles with the local and foreign elite. He's still a middleman; he uses his connections and powers of persuasion to urge landowners and tribal chiefs to sell grain and herds to feed the occupiers. Talking to Zinger – to whom he announces his decision not to sell the occupiers his produce – Yusuf feels the humiliation of a nation that has capitulated without fighting, 'without tasting either triumph or honourable defeat'.

There is nationalism in Daneshvar's writing, though patriotism may be a better word. (Even that may be difficult to sustain; patriotism requires a *patria*, and that was a chimera for many Iranians in 1940.) Rather than the robust rewritten history of less subtle nationalists, Daneshvar uses a luscious nostalgia to illuminate Persian feeling.

To read *Soo va Shoon* is to take sociable refuge in a big house whose roof projects from all sides with wide shade-offering eaves. It is water gurgling in channels through a garden shaded by the sour cherry tree and that cynosure of feminine beauty, the Shiraz cypress. After a long afternoon nap, there will be the murmur of visiting relatives or friends from the tribes upcountry; their fluid entrances and exits will be an excuse for opium, wine and conversation with due homage to the poets and aphorists. Shiraz, remember, is the place of Hafez, that strenuous oenophile and Muslim imagist; a group of devotees gather weekly at his tomb to recite his verses and to pour libations on his stone lest they be accused of meanness.

Reading *Soo va Shoon*, you think: perhaps it never existed. Look at the Islamic Republic around you – anti-sensual, distrustful of secular pleasure. Bita, who was very young when *Soo va Shoon* was published, will correct you; it did exist. Her summers in Isfahan were spent in not dissimilar surroundings. From her own recollections, she summons passages that are reminiscent of the novel. And yet, by the 1970s, the old life was going – with the break-up of estates, the effects of the Shah's land reforms and modernity's more opaque menaces. Living off revenue from distant holdings; concerned, with noble bleeding hearts, for the welfare of tenant farmers; Yusuf and Zari inhabit a world that is, no less than a Tolstoy estate or a cherry orchard washed by the Volga, on the verge of extinction.

But Bita thinks of other things when I ask her about *Soo va Shoon*. They overshadow even the narrative, which describes Yusuf's attempts to resist the British and their local collaborators, and climaxes with his murder at the hands of these enemies. For Bita, *Soo va Shoon* is a book of women.

The theory of their lives is a cushioned supporting role, punctuated by pregnancies and embroidered with sensual joys. But the first part of this theory runs true for none of the novel's three main female characters. At least Zari, of the three, has fourteen years, the span of her marriage to Yusuf, before it is subverted, and she gives every impression for that period of appreciating her good fortune. Eventually, however, all three must react to the bad luck, ineptness and (sometimes) callous boorishness of their men. But the sensuality remains – palliative, perhaps, or defiant?

Fatemeh, Zari's sister-in-law, was broken years ago by her father's love affair with Sudabeh, an Indian dancer. The humiliation was too much for Fatemeh's mother, who took refuge in the shadow of the shrine of the Imam Hossein, at Karbala in Iraq. (There, she ended her life in penury, reduced to working as a lady's maid and ignoring letters from her children in Iran.) Fatemeh lost her six-year-old son to illness. She then lost her husband; seized by an obscure rage, he galloped into one of the stout pillars that hold up the British consulate in Shiraz (the 'consul's daughters', people call these pillars, with scant regard for slim English waists). Now, Fatemeh survives on opium and the longing, a doleful echo of her mother's, to die alongside the Imam Hossein – the only man, perhaps, on whom a woman can rely.

'Every Iranian,' Bita says, 'has an Aunt Fatemeh' – tragic, reliant on her extended family to jolly along her solitude. Fatemeh is awed by God's ability to send down unhappiness and desolation, but she has a winning propensity to tear strips off her other brother, the collaborator Khan Kaka. (Blind servility is out for these thinking Iranian women.) And she gives a description, remarkable for its lack of rancour, of the dancing Sudabeh – the author, it could be argued, of her misfortune:

> She had a black mole on her upper lip. She was dark. She really wasn't all that beautiful. She was short. She had big dark eyes and long hair. When she wasn't smiling, she looked like an owl. But when she smiled . . . flowers poured from her mouth. Everyone, men and women, stood and clapped, and it was as if she were naked, wearing only jewellery. But she wasn't naked. She had a brassiere covered in jewels and a kind of stocking over

her whole body ... she made every part of her body move. Not only her shoulders and tummy and eyes and eyebrows, which is easy enough, but also her chin and nose and ears and pupils. It was as if she was dancing on the corpse of a man ...

The sensuality is most evident in a bathhouse that belongs to Izzet ud-Dawleh, the third of Daneshvar's important female characters. (Her honorific, 'Prestige of the State', no doubt derives from her late father's position as a magistrate, and such absurd and self-regarding sobriquets were common in those days.) There are spurting fountains, white marble basins and a serving girl bearing cold lemon juice in a china cup adorned with the old Persian image of the hen and rose. There is Izzet ud-Dawleh's disclosure of the ingredients of a secret compound that makes her hair lustrous and strong: ground coffee, henna, cocoa and essence of camomile. (Having rubbed this into your hair, seal with leaves from the walnut tree and leave for a day and a night. The coffee prevents the henna from dyeing your hair red.) There is the division between what is fit for men's eyes and what is not.

In *Soo va Shoon* there's a sense of exasperation at the men, and also of social need – for who in Iran would dispute that marriage and child rearing are the sole feminine stations? (Women are parcels, to be passed from father to husband, sealed and tied with ribbon to prove that they haven't been opened before.) But Daneshvar doesn't make the mistake of creating paragons instead of women. (Except for Zari, with whom the reader falls blandly in love.) Izzet ud-Dawleh is the best example. This cynically mawkish creature with her acquisitive squint pollutes society as much as she is a victim of it.

No one disputes that Izzet ud-Dawleh had it rough. In the early days of her marriage, her husband ordered her to serve alcohol to the prostitutes he entertained. Now, widowed and living with her lecherous son Hamid, her financial position is precarious; to maintain a respectable façade, she dabbles in the smuggling of food, and even arms.

She once took in a destitute named Ferdows, whom she found in the street. (She claims to have taken in the girl for philanthropic reasons, but she must also have liked the idea of having a slave.) Izzet ud-Dawleh –

who relates the story to Fatemeh, in brutal street Persian – hadn't reckoned with her own son and husband. 'It didn't take more than a week,' she says, 'for one of those two, either father or son, to knock up that will o' the wisp.'

Scandal loomed – how to explain the parenthood of the child? Izzet ud-Dawleh tried to persuade Ferdows to marry the concierge – the one 'whose mother would go once a month to Jew Town and buy him a girlie for three *tomans* and doll her up in fake satin and bring her home, and return her to her owner after the satin had been ripped . . .' For three days, Izzet ud-Dawleh imprisoned Ferdows without food or water. Still, the girl threatened to file a complaint; she supposed, Izzet ud-Dawleh recalls, that 'she could profit from the bastard in her stomach'. Izzet ud-Dawleh put her right. 'I thrashed her to kingdom come, and thank God she started bleeding . . . she calmed down once the brat had been buried.'

Despite all this poison, Bita harbours a kind of affection for Izzet ud-Dawleh. In Bita's eyes, she's not a monster but a survivor.

Izzet ud-Dawleh makes sure she is chief mourner after Yusuf's murder (and chief opportunist, for his death revives her former designs on Zari as a prospective daughter-in-law). Short of her thirtieth birthday, getting bigger with her fourth child, Zari recognizes the futility of her remaining apolitical: 'I wanted to raise my children on love, and surrounded by calm. Now I shall raise them on hate.'

Soo va Shoon ends with death and honour. In that way, every Iranian reader knows, it recalls the martyrdom of the Imam Hossein and Siyavash's heroic death. Processing through the streets, the mourners for Yusuf are held up by policemen and soldiers, who order them to go back to work. (A citywide strike has been called in protest at the murder.) There is a standoff, punctuated by threats – the traders will have their licences revoked – and protestations of peaceful intent. Imagine this is the battlefield at Karbala, one of the mourners says to the police captain facing them. Are you prepared to be Shemr?

The security forces attack with their rifle butts; shortly after, there's the sound of gunshots. For a moment, Zari and Khan Kaka are left alone with the coffin, which lies in the road while the dust flies around them. Having found people to help them take the coffin back to the house, they find the garden with the sour cherry trees and the cypresses full of wounded people.

The Persian women supplied the answer. Out from their walled courtyards and harems marched three hundred of that weak sex, with the flush of undying determination in their cheeks. They were clad in their plain black robes with the white nets of their veils dropped over their faces. Many held pistols under their skirts or in the folds of their sleeves. Straight to the [parliament] they went, and, gathered there, demanded of the President [of the parliament] that he admit them all. What the grave deputies of the land of the Lion and the Sun may have thought at this strange visitation is not recorded. The President consented to receive a delegation of them. In his reception-hall they confronted him, and lest he and his colleagues should doubt their meaning, these cloistered Persian mothers, wives and daughters exhibited threateningly their revolvers, tore aside their veils, and confessed their decision to kill their own husbands and sons, and leave behind their own dead bodies if the Deputies wavered in their duty to uphold the liberty and dignity of the Persian people and nation

W. Morgan Schuster, *The Strangling of Persia*, 1912

The woman wore a black coat and headscarf, which accentuated the paleness of her skin. She was taller than average: not slender but big-boned rather than stout. Her face was long, with a fulsome jaw and a handsome Iranian nose with a bump on the bridge. It was 22 November 2002.

The men in the hall and the outside courtyard were waiting for her to speak; they were seated on chairs or cross-legged on the floor, or standing in cramped huddles wherever there was space. Earlier, as we converged on the hall under a winter sun, I'd noticed that many of the men had grey skin but that their eyes were bright. Their brightness was a reflection, I realized, of her brightness.

Until four years ago, many of them had known her as the daughter of Darioush and Parvaneh Forouhar. She was an artist with two children and an ex-husband, and she lived in Germany. Then, on 22 November 1998, her parents had been murdered by Iranian security agents, and she'd lifted the burden of their death in a way that had surprised people. Now, her parents' friends and supporters looked at her admiringly. There was something clean about her; you couldn't impugn her motives as you could those of a politician.

She spoke. 'Four years have elapsed since the savage killing of my father and mother, Darioush and Parvaneh Forouhar. One black evening in exile, my ears rang with the cries of a friend, portending murder. That evening, my sons sobbed and my brother slammed his head and fists against the wall, and every sound that reached our ears mingled with weeping and curses.'

Her voice was strong but slightly hoarse – the hoarseness that comes from omitting to project, actor-like, from the diaphragm. The lines were cadent, with a preference for pure Persian over Arabic loan words. She'd brought absolute silence to the prayer hall and the courtyard. Through an open door behind her, I could see women gathered in a smaller courtyard, looking intently in different directions as people do when listening to a voice from a loudspeaker.

'Three days later, collecting their lifeless bodies from the Tehran coroner's office, I refused to sign until I'd seen their wounds. After a brief dispute, the blanket was lifted to allow a daughter to see the mutilated corpses of her father and mother. They didn't allow me the time to weep or kiss their wounds.

'Seven days later, they vacated my parents' house. They'd occupied it since the bodies' discovery, under the pretence of searching for traces of the killers. Even the house had been branded with hate. It was as if everything had been caught up in a whirlpool of wildness, and swallowed. Their notes, their political documents, their life; they had been pillaged.'

In the smaller courtyard, the women wiped away tears. Here, the old men were blinking fast. Young men, too young to have known the Forouhars for very long, looked determinedly at the ceiling. Parastu Forouhar told us that her parents had been persecuted not only by the Islamic Republic but also by the Shah; that the judicial file prepared in the case of their murder and that of two other dissidents had been distorted, penalizing the men who had carried out the killings without seeking those who had issued the orders; that the brave family lawyer, Nasser Zarafshan, had been jailed for trying to find out more about the murders. She repeated this common knowledge in such a way that we sympathized with her private grief and felt collectively affronted.

'I, a child who knew my father's broad frame from visiting hours in narrow cells; I, who caressed the scars on his head . . . the gashes left by the knives of security agents . . . who was enamoured of his eyes' electric defiance and its corollary, the refuge and support that he gave to the oppressed; I, who lovingly observed my mother's fortitude and constancy over a hard cycle of years, who witnessed unstinting life affirmation in her tender eyes and unaffected words; now, four years after their torn bodies joined with the earth of their beloved motherland, I ask you . . . how do they define that word justice in this country of ours?'

There's something admirable about a man who, having related the machinery of oppression lined against him, asserts that he won't give in. But when a woman does this, and she is the subject of a state that appreciates women only as long as they exercise a demure faith, stand up to malefactors against the faith and raise fine lads and girls in its name – in short, when that woman steps from the cramped hut that has been prepared for her – it becomes inspiring.

Parastu Forouhar answered her own question: 'Today, the meaning of justice lies in the pledge of steadfastness that we make, each of us . . .' She and the families of the other two victims would request that the UN investigate the case. She appealed for supportive signatures. 'The victims of these murders aren't only connected to us, but to every Iranian.'

Then she was gone, and with her the sense of security that came from being dozens thick in a warm room. Looking around, it was possible to make out the informants and agents among us, not bothering to hide their identity – scowling through dank beards, pocketing notes made on a pad. There was no telling what institution they belonged to. The Intelligence Ministry? The police's intelligence wing? The public prosecutor's office?

The men in the hall were moving slowly, out of the door and into the courtyard, up some stairs and into the street. The younger ones might be students. Perhaps they'd been involved in some recent campus demonstrations; if so, their presence illustrated less their admiration for Forouhar and his wife than a desire to make a political point. As the crowd bottle-necked, slogans were raised: calls for the release of political prisoners, homage to Forouhar and his mentor Mossadegh. I squeezed through the door, into the courtyard. Everyone was looking up. There was a man on the roof, filming us for some sinister archive. 'Spy! Get lost!' the people chanted, and he slunk away.

Upon exiting the prayer hall, I found out later, Parastu Forouhar and the lads with her had been attacked by men wearing black leather jackets, using belts and knuckledusters. An impromptu bodyguard had formed around her, absorbing most of the attackers' punches while she walked quickly away. Some of the lads were arrested by the men in leather jackets; these men didn't bother to introduce themselves, let alone show a warrant.

I came out into the alley at the very moment that a uniformed policeman brought down his truncheon very hard on the head of a young man. The man clutched his head and the other people stared. Suddenly, a woman screamed at the policeman: '*Kesafat!*' Filth! The woman was middle aged and purple with rage, and there was a short, uneasy man by her side. The young man dropped to his knees. Again, the woman screamed: '*Kesafat!*' The policeman was astonished, but shame-faced as well. The short man placed his hand underneath the woman's elbow, but his soothing words only served to make her angrier. She shrugged him off, saying through clenched teeth: '*Don't* touch me!' As she walked away, she muttered for the last time, '*Kesafat*', and shook her head sadly. The policeman looked at the nervous man, following after, as if to say: 'Can't you control your wife?' Then he took the boy away.

The alley was full. The people started walking towards the main road; many of them wanted to sign Parastu Forouhar's letter to the UN. The women linked arms. A group of young men walked ahead. They raised slogans. 'Death to the Taliban, whether in Kabul or Tehran!' I tried to look inconspicuous as they shouted, 'Death to Britain!' an allusion to the belief that the British engineered the mullahs' rise to power.

As we watched, curious people gathered on the balconies and roofs to watch. The marchers shouted, 'People! Why are you sitting down!' It had been a common slogan at the time of the Revolution. The wittiest slogan was deployed against Rafsanjani. They made fun of his inability to grow a beard and told him to get back to his pistachio farm. Then we reached the main road. At one end, there were hundreds of policemen, with the plainclothes thugs not far off. The young men marched as far as the first line of police, while the thugs tried to get through. Near me, a woman turned to her companion and said, 'I'm scared, but something has to be done.' That evening, we heard that some of the protestors had been badly beaten.

❖ ❖ ❖

A few evenings later I was standing in a small alley outside a white gate. A minute or so after ringing the bell, I heard shuffling on the other side and the gate opened. I recognized the servant standing there from an earlier visit, but there was no sign of the poodle that had greeted me the last time. The poodle, I remembered, had been the sole witness to the killings of Darioush Forouhar and his wife. She had passed away, the servant told me, of old age.

He escorted me across a short driveway and up two or three steps, into a brightly lit hall. To someone like me who hadn't known it before the murders, the house seemed warm and inhabited. There was the sound of voices in the next room. The telephone on the hall table rang. You half expected Darioush Forouhar to emerge from his small office leading off the hall, or Parvaneh Forouhar to come downstairs and offer you a cup of tea. I glanced in at the door of Forouhar's office. It was dark.

That was where a group of his colleagues from his small Nation of Iran Party had found him, sitting in the chair nearest the door, his hands hanging by his sides and his mouth open. Later, during the rush of interrogations, leaks and rumours, they discovered that, before stabbing him eleven times, his killers had orientated him towards Mecca. That's what Muslim butchers do when slitting an animal's throat. (They also stipulate mentally their intentions to kill the animal according to God's law.)

They had found Parvaneh in an upstairs room. She was lying on the floor, having been stabbed twenty-four times, and she was bruised around the neck.

After making the discoveries, the men had stood at the foot of the stairs, trembling and wondering what to do, when the telephone on the hall table rang. One of the men went to answer it, but the line went dead when he put his ear to the receiver.

My reverie was disturbed by the servant, calling Parastu to the telephone. She came out of a large L-shaped room opposite her father's office, accompanied by a visitor, a young man I thought I recognized from the prayer hall a few days before. The man told Parastu that he would expect her answer to his question, and left. Parastu invited me to go into the L-shaped room and have a cream cake. She picked up the receiver and started speaking in German.

From the doorway of the L-shaped room I looked at the big portrait of Mossadegh, whose death in 1967 under house arrest is one of Parastu's early childhood memories. (She was five years old.) I had a feeling that the portrait had been there a long time; I had a similar sense of longevity looking at some of the other things in the room – pictures hanging on the wall; slightly frayed furniture and carpets; inexpensive bibelots. On a low table in front of some armchairs there was an ashtray, half full of butts, and two glasses containing the dregs of tea. There were cakes, lying in a box emblazoned with the name of a patisserie. I helped myself.

Eating my second *rollette*, I realized there was an odd coexistence in the L-shaped room, between the permanent seeming things and other objects that post-dated the murders. On the far wall there was a pompous portrait of Parvaneh Forouhar, draped in an Iranian flag. The wall space around her had been covered in framed photographs of her and her husband, many of them dating to their handsome early marriage. These weren't an accretion of mementoes – rather, a conscious regiment, overpowering and stifling. Two chairs near the window overlooking the driveway had been draped with Iranian flags. I contemplated a third *rollette* and Parastu came back into the room.

We'd met through a mutual friend. We had other acquaintances in common; Parastu had come briefly to our house. In different circumstances it's possible that we'd have become friends, but I had only seen her in connection with her parents' murder. Even on the rare occasions that we'd met socially in the house of the mutual friend, it had been impossible to forget the murders. Parastu came to Iran for defined and declared reasons:

to see lawyers, to try and goad the authorities into reopening her parents' case, to maintain links with her parents' friends and colleagues. Her sons and boyfriend stayed in Frankfurt.

We were sitting with fresh tea brought in by the servant. Parastu lit a cigarette and we discussed Bita's expanding tummy. My wife was in the middle of her second trimester; she was attracting stares and comments in the street. Beggar women, I told Parastu, tried to curry favour by expressing the hope that, *inshallah*, Bita would be granted a son. When Bita replied with choice insults, they would retort: 'I hope you have a daughter!' Parastu laughed, rich and loud, and I laughed with her. Then I remembered where we were and what had happened here. I thought of the strength you would need to stay in the house where your parents were murdered.

Later, reading a childhood memoir she'd written, I guessed that there hadn't been much question of her staying anywhere else. Forouhar had spent a total of sixteen years in the Shah's jails, and one as a prisoner of the Islamic Republic. He was the kind of man who, having being forced by jailers to stay on his feet for forty-eight hours, responded to an invitation to sit by saying, 'I prefer to stand.' His wife once promised the Shah's prime minister that she would immolate herself publicly if he didn't tell her where her husband was being detained. She saw Forouhar off to jail not with tears but a rendition of a nationalist anthem. Parastu wouldn't be hounded out. She would carry on laughing.

The laugh is the first thing people notice when they meet her. Given the history, and the setting, you might regard it as out of place. But humour was essential to the Forouhars – it was a way to raise spirits and make fun of tyrants. Dictatorships are vulnerable to ridicule. Their most serious people lay down impenetrable treatises that everyone is meant to read and no one does. Their personages are protected by laws that have been passed ostensibly to guard the nation's dignity, but are actuated, in fact, by a terror of mockery. And the enforcers, the agents of repression and internal security; you might think of them as buffoons, ignorant of when they are sinned against, and when they are sinning.

Parastu laughs as she describes meetings with the police to learn what will and will not be tolerated during the annual commemoration of her parents' death; she laughs while relating visits to obscure courts where she is treated with exceptional coarseness by judges wearing slippers. It's in the genes. Her father would greet the agents tapping his telephone with the salutation, 'Salaam to all our listeners'. He rocked with mirth having found an imperial agent searching his fridge for pamphlets. Once, having left a political discussion in the L-shaped room to tune into a foreign news channel, he chanced upon a frequency that was broadcasting the conversation. The snoopers had got their wires mixed and it became a family joke.

Her parents and her idea of Iran are intertwined. This sense must have intensified now that the Forouhars are part of the soil. (Much as they loathed him, they must have understood the Shah, clutching his clod as he fled into exile.) Speaking about Iran, Parastu uses a word that means feeling, hess, in a way that is both breathy and sibilant, stretched over an exhalation of cigarette smoke. Iran is a hess, an ideal, a tragedy.

It's jails and the opposite of jails – the mountains outside Tehran, where her father took her when he was free. Like Zari's in Soo va Shoon, a book that she also admires, Parastu's nostalgia lies in non-Islamic Iran. Her Iran is not Hossein but the night of Yalda, the longest of the year, celebrated with pomegranates and the poems of Hafez; it's the thirteenth day of the (pre-Islamic) Persian New Year, when unmarried girls tie a knot in the grass and wish for a family. When she thinks of pilgrimage places, Qom and Mashhad are absent. Hers is secular – Ahmadabad, the resting place of Mossadegh.

In 1978 she and her friends threw on scruffy clothes and headscarves and went to south Tehran to get their kicks from a dynamic proletariat. They sprinted away from tear gas, distributed tracts. In 1979, she was entranced by Khomeini and a little embarrassed when her father accepted a job in a ministry as unrevolutionary as Bazargan's. When the War started, she volunteered to be a field nurse, but fell ill, which made her worse than useless, a burden; she was evacuated in an ambulance meant for injured men.

In Tehran, things were changing. The universities were closed. Independent voices were unwelcome. Standing on street corners, selling newspapers that were sympathetic to Bani-Sadr's government, Parastu was beaten up by Khomeinists. It was the beginning of her retreat from politics. 'I wanted as much as possible to hide myself from the gaze of others, or to stay at home. I remember trying not to look at anything or anyone.' After Bani-Sadr's flight, her father went underground, a fugitive from the regime he had served.

He emerged briefly to bless his daughter's wedding, in a secret place, to a young activist. Homayun was an ally of Forouhar's; he carried on daubing slogans even when daubing slogans could get you tortured or killed. He glowed with integrity next to his father, a general and military judge whom Parastu describes as 'taking on the colour of whichever government happened to be in power'. (The general boycotted the wedding.) Parastu was attracted by Homayun's political fire; her childhood had been warmed by such passions. But she no longer shared them.

She'd shown artistic aptitude at school. Now she had sad, negative motivations to take up classes: to isolate herself from the ugliness around her, to ignore it – not to confront it. (These motivations had also influenced her decision to marry, at an age her parents considered to be too young.). She wanted, she says, a place that was 'safe' and 'healthy'.

You paint a flower. You have the satisfaction of knowing that, even if the flower can be destroyed, the fact of its creation cannot. Parastu enjoyed draughtsmanship – clean lines, taking pains over a leaf or a glass. The effort of concentration distracted her from new realities: her father's surrender to the authorities and his disappearance, behind bars, for months on end.

In time, the universities reopened and Parastu began a fine arts degree. Looking at the photographs you're surprised by smiling students. But this isn't defiant Forouhar mirth; it's self-deluding, blocking. Before, the kids in the photographs were Communists, Mujahedin, Liberals. Now, they want to forget. No one asks for or expresses an opinion about politics. They don't follow the War, even though the faculty encourages them to do so – and to devise ways of glorifying it. 'For us,' Parastu says, 'it was more interesting to set up a screening of films by Wim Wenders, to meet for life drawing in the home of a friend, to walk in the mountains.'

While boys threw themselves to their deaths, Parastu and her friends examined the way that a breast hangs from a girl lying down. They travelled outside the city, to be in places without slogans. They tried, as unobtrusively as possible, to remove the prism that the Revolution had transposed between them and their own actions. About that time, very unexpectedly, her father was released from prison. The War, and the need for national unity, convinced him to soft-pedal on politics. Parastu concentrated on art.

She was lucky to have an independent-minded teacher. He gave a course on the Iranian miniature – not the subject matter, which is the glory of kings, but technical themes, line, composition, jewel-like colour and conventions. Even that must have been objectionable in the eyes of those who were jealous of the past, and wanted to eradicate it. (How else to explain Khalkhali's alleged suggestion, never successfully acted upon, that the ruins at Persepolis be bulldozered?) Some of the students, Parastu included, had pieces in an exhibition to celebrate the one thousandth anniversary of the *Shahnameh*. By then, the enlightened teacher had gone; he was retired as soon as some Khomeinists among his students got their MAs; now, they were qualified to replace him.

'When I look back on what we did at university, I don't think it was art. Art can't be produced in such an environment. You have to have the opportunity to think freely, and we didn't have that freedom. Everything was closed, everything was channelled.'

'Is there nothing you remember that carries artistic value?'

'The feeling.' The *hess*. 'The feeling of that instant. A direct communication between the created being and the world around.' She loved doing quick portraits – the movement of the brush was more important than the likeness.

'I thought they didn't allow that feeling to emerge.'

'No, they didn't allow art to happen. Art is the stage after this feeling.'

There was another feeling, even sweeter. It was a delightful pregnancy – because of the conditions, not in spite of them. Choosing colours; sewing baby-sized sheets; these things gave her a clean feeling. She gave birth to a Kourosh, whom the Greeks called Cyrus – a pre-Islamic, nationalistic name.

But her marriage to Homayun had soured. 'Homayun wasn't prepared to give up politics. I became for him a sort of individualist "intellectual" who was pursuing this art, something he didn't believe in. He would say,

"You've abandoned your old ideals, the ones I came to know when I first met you." From then, our mental worlds diverged.'

He left first, for Germany, and she followed two years later – so that he could see the children and they could divorce. The War came to an end. Then Khomeini was dead. She saw it on TV – bodies covering his coffin like bees. In Tehran, Forouhar was getting involved in politics again. Gradually, she began to enjoy her new life.

❖ ❖ ❖

Iranian Studies was an incidental and unwelcome part of my degree course. I studied it in a distant and lackadaisical way. I had few opinions about contemporary Iran. Rafsanjani was the president. He was a 'pragmatist'. His aim was to 'reconstruct' the country. That sounded dull, but I was unlikely to see Iran anyway. Tense bilateral relations made it hard for British students to spend time at an Iranian university. I was more interested in my other field of study, India. After graduating, I went to live there. Later, I moved to Turkey. A few years on, someone asked me what the Persian for 'yes' was, and I realized that I'd forgotten.

In 1999, I visited Iran from Istanbul. Within three weeks, I had met, fallen in love with and decided to marry Bita. (Getting her from stage one to stage three would take a year's lobbying.) I had a personal interest in learning about Iran.

I had an impression of transition. Most people liked Muhammad Khatami, who had taken over from Rafsanjani as president. Khatami wore a silver beard and smiled a lot, but not too much. He was a *seyyed*, but women attested to his ability as a flirt. His reformism seemed to mean rounding the harsh edges of the Islamic Republic, and trying to reconcile Islam and democracy. When he addressed crowds of young people, girls would scream themselves silly and faint.

Everyone said that reform was linked to newspapers. Under Rafsanjani, I learned, there were ten, most of them turgidly conservative, and the kiosks wore a deserted aspect. But Khatami had overseen a dramatic increase in the number of dailies and weeklies, and now there were crowds around the kiosks. The various titles were laid out on the ground; people would survey the columns, looking over the shoulder of the person in front, and argue about what they meant. The combined circulation of the newspapers and magazines was close to three million, twice the figure of two years before. Editorialists expressed their views with a vigour that Iran hadn't known since Bani-Sadr. It was hard to imagine there had ever been silence.

That first trip to Iran, I spent a lot of time on Bita's doorstep. (I was early for dates.) I would bring a newspaper with me and try and decipher it. These papers were attractive and manageable, appearing on *Le Monde*-sized newsprint. They had colour photographs – of an Iranian actress, perhaps, accepting a prize at a festival abroad – and headlines that often ended with a question mark. They contrasted with the dark, heavy broadsheets that were produced by the opponents of reform, the conservatives. I remember trying to read one of these newspapers, *Kayhan*, and getting my hands inky from a photograph of some Israeli soldiers shooting at a Palestinian boy. You never saw smiling people in *Kayhan*. The headlines often accused America – along with their lackeys, Israel and Britain – of plotting against the Islamic Republic.

While I was in Tehran, I met Mohajerani, Khatami's minister of culture. The conservatives loathed Mohajerani because his ministry issued the news-paper permits. His people had authorized the first pop music releases since the Revolution. They had permitted (after a four-year wait) the screening of a film that upset conservatives because it told the story of a young woman trapped in a traditional marriage. A dozen publishing houses had been set up to print the books that Mohajerani unbanned. I found the minister arrogant and supercilious. Bita, who accompanied me to translate, found him arrogant and dishy.

In 2000, I moved to Tehran. That February, Iranians had elected a reformist parliament that was opposed by an appointed upper house, the Council of Guardians, which was dominated by conservatives. Both groups contained mullahs, but the handsome ones were mostly reformists: Khat-ami, for instance, and a friend of his, a theologian called Mohsen Kadivar. (Quite a few clerics, from both sides, were lumpy and uncouth; during an interview, one used my calling card as a toothpick.) Lay politicians of both persuasions inclined to stubble and greasy suits.

In time, I discovered ideological differences between the two groups.

Reformists were less suspicious of foreigners and women. They emphasized the enlightened and democratic parts of the constitution. Conservatives promoted repressive bits of the same document – bits that emphasized the unaccountability of the most senior clerics. Both groups claimed to represent the Islamic Republic. So did the man in the middle, Rafsanjani, who still had a lot of power; he'd thrown his weight behind Khatami's election, but had since got closer to the conservatives. I wondered how the Islamic Republic had fractured in this way.

It started, as family feuds do, with the death of the patriarch. Old Khomeini should have realized that his odd suit of clothes, the Guardianship of the Jurist, fitted none but him. (That it fitted him, furthermore, was in no way down to the quality of its design, but rather his own pre-eminence in politics and theology.) Khomeini should have taken the opportunity, with Montazeri's dismissal, to trim the powers that had been given to that Guardian, who was called the Guide, and to make the office an elected one. Instead, a few weeks before his death, he convened a constitutional assembly that did the opposite.

The assembly increased the Guide's formal powers and described his authority, with an explicitness that was absent from the 1979 constitution, as 'absolute'. It took note of the Imam's assertion that the Guide could take any decision he deemed to be in 'the interests of Islam and the country'. For Khomeini, with his quasi-divine charisma, there had been no need for explicitness. His utterances were regarded by millions as religious ordinances. But the next Guide wouldn't have these advantages. You can't force people to believe that, simply because the representative of God has died, the man now sitting in his chair is the new representative of God.

Montazeri was out of the running. No object of emulation was suitable for the post. (The rest were old, apolitical or sceptical about the religious legitimacy of the Guardianship of the Jurist.) In line with Khomeini's wishes, the constitutional assembly removed the stipulation that the Guide must be an object of emulation. That meant that hundreds of lesser ayatollahs were eligible, provided they had the necessary piety, courage and 'good managerial skills'. Having been the preserve of a revered divine, the Guardianship of the Jurist could be conferred on a politician in a turban.

The day after Khomeini's death, the Assembly of Experts (the body that chooses the Guide, and assesses his performance) elected Ali Khamenei, who had been president for much of the 1980s, to be the Imam's successor. The new man was middling in theological terms – even allowing for his abrupt elevation to the rank of ayatollah. Many people were uncomfortable with the idea that someone they had elected, knowing they could oust him, to the mundane office of the presidency, had become the vice-regent of God. The new Guide would need to manoeuvre with and against other people and institutions – Rafsanjani, who would go on to replace him as president; parliament; appointed bodies like the Council of Guardians. He would be like any other politician.

These ironies were not lost on the men who had raised the Revolution and fought the War: Alavi Tabar, for instance. He joined a think-tank, the Centre for Strategic Research. Its members were concerned by the gap that had emerged between the aspirations of normal Iranians and the stagnant lives they were obliged to live. They studied subjects like poverty and modernity and the characteristics of a market-driven economy. They asked if a more flexible approach to the interpretation of religious texts might rescue Iran from its morally superior poverty.

Their favourite philosopher was Abdolkarim Soroush. Shortly after the Revolution, Soroush had been involved in the purging of the universities – 'under the control,' he said, 'of my passions'. Now his furies had dropped. He had conciliatory views on the influence of Western culture and norms. He distinguished between the essence of knowledge and its forms; Iran, he argued, must engage fully with the former, even if it was found in the West. Human interpretations of Islamic law, he said, were endlessly open to revision. Later, he would argue for the detachment of the clergy from mundane interests; such a detachment would benefit religion. The traditionalists, or conservatives, shook their heads; was this child of the Revolution making the case for secularism?

At that time Muhammad Khatami was Rafsanjani's culture minister and he was a cautious ally to the progressives. War, the old justification for harshness and repression, was over; Khatami saw no reason to prolong the harshness and repression. He and his deputies breathed life into Iranian cinema, encouraging filmmakers to be critical and socially acute. His ministry allowed *Salaam*, a relatively progressive newspaper, to start operating, as well as magazines that discussed women's rights – or rather the lack of them. The minister was an interesting paradox. While acknowledging the influence of Western philosophy on his understanding of liberty, he regarded the West, monolithically, as 'our philosophical and moral opposite'. Even so, he argued, it was desirable to have an exchange of ideas.

A fourth pillar emerged, the so-called 'radicals'. Former Khomeinists, they had been orphaned. They opposed what they regarded as Rafsanjani's attempts, under the guise of economic liberalization, to undermine the Revolution's commitment to social justice. They opposed his modest efforts to improve relations with countries, like Germany and France, which had supported Saddam during the War. Some of them regretted Khamenei's appointment as Guide; they made a controversial trip to Qom to see the ostracized Ayatollah Montazeri. With the aim of restricting Khamenei, some of them declared the Imam's last will and testament to be the only acceptable template for Islamic government.

The radicals might have got away with antagonizing either Khamenei or Rafsanjani, but not both. Rafsanjani did not oppose a successful push, by Khamenei's allies, to expand the interpretation of the powers enjoyed by the Council of Guardians, which made up of the Guide's direct and indirect appointees. Henceforth, the Council was empowered to vet candidates for election; before the 1992 parliamentary polls, they disqualified nearly a third of all candidates, including dozens of sitting deputies and hundreds of radicals. In their anger, the radicals resorted to an unfamiliar, Western lexicon. Their 'rights' had been trampled underfoot. The Council of Guardians had behaved in an 'autocratic' manner. More 'democracy' was the answer.

Rafsanjani might have benefited from the radicals' elimination, but the conservatives who took their place were at best lukewarm about economic reform, and they were increasingly gathered around Khamenei. They and the Guide had a common aversion to Western cultural 'contamination', which they believed Rafsanjani wasn't doing enough to stem. Khatami was issuing permits for the publication of books, including feminist novels, which had previously been considered beyond the pale. Rather than mosques and prayer halls, Karbaschi, whom Rafsanjani had appointed to be mayor of Tehran, built cultural centres and laid out parks.

In 1992, the Guide forced Khatami's resignation, and the following year, that of the head of the state broadcasting corporation, who happened to be Rafsanjani's brother. (The latter had permitted such outrages as the broadcast of cartoons showing boys and girls having adventures, and an animation of *Around the World in Eighty Days*, whose virtuous hero was 'explicitly introduced as English'.) Rafsanjani kept Khatami on as his culture adviser, and let him thrive as the head of the National Library. But he was forced to appoint a hardliner in Khatami's place, and the permits dried up.

Gradually the Guide widened his base of support. Recruitment was stepped up to paramilitary groups like Hezbollah and the Basij. (Basiji reservists were rewarded with shorter periods of military service, and university places.) Encouraged to feel loyalty specifically to Khamenei, these groups responded to his call that they identify the 'enemy' that was 'waging an onslaught against the clerical system'. They trashed the offices of progressive magazines and intimidated Soroush and his supporters. (This campaign led to the suspension of Soroush's classes at Tehran University; he went abroad for an extended sabbatical.) In Qom, the special clergy court, whose chief answered directly to Khamenei, was harassing and handing out prison terms to Montazeri's supporters.

The Guide, formerly shy, now seemed to approve of the sycophancies being laid at his feet. 'God,' the head of the judiciary observed, 'has ranked obedience to the vice-regent equal to obedience to God and the Prophet.' An obsequious deputy suggested that parliament content itself with accepting 'the commands of the exalted Guide unconditionally'. It was an odd turn of events, for Khamenei had previously favoured a restricted reading of the Guide's powers.

Rafsanjani didn't succeed in shifting capital from the public to the private sector; he blurred the distinction between the two. You got businessmen being approached by senior security officials who told them: 'Look – import anything you like, except booze, and we'll ensure fast-track service through customs and even the use of our jetties on the Persian Gulf.' You got items whose import was officially a government monopoly being brought in by officials acting in a private capacity. These people didn't have conflicts of interest. They had dovetailing interests.

None more than Rafsanjani. During the War, he had overseen the setting up of a network of private universities that greatly expaned in the 1990s, a vastly lucrative monopoly that he retains to this day. The president set up a mechanism for controlling the sale and export of pistachios – of which he was a big producer. Sometimes, cracks showed – establishment figures might have been involved in huge corruption cases that the courts only half revealed – but the rottenness spread. There was a revealing contretemps at a banquet between a female member of Rafsanjani's family and the widow of the former prime minister, Rajai. Where, the distressed widow demanded as she surveyed platters of food piled high, was the old revolutionary concern for the oppressed?

The people cottoned on. A bed in a full hospital ward; a comfy posting for your son during his military service; a favourable court ruling in a dispute with your business partner; a blind eye to an indiscretion, drunk, with a woman other than your wife: these privileges were commodities. Commercial decisions depended not on what you were offering, and at what price, but whom you knew and what you paid under the counter. People remembered Rajai behind the wheel of his Paykan, driving to work at the Prime Ministry. Now, you found convoys of Mercedes and outriders screeching through the capital, with bodyguards leaning out of windows, waving the riffraff out of the way. You found mullahs who were known for Anglophobic Friday sermons queuing at the British Airways check-in, London-bound for a medical check-up and a spot of shopping.

Later, when he was forced to defend his presidency, Rafsanjani portrayed himself as a friend of democracy. He'd achieved office 'through the votes of the people'. It was his special 'pride' that the people had wanted him in the speaker's chair, before raising him to the presidency in 1989, with a crushing majority of the votes cast. He drew attention to his position on the Assembly of Experts, which is elected directly by the people.

In the immediate post-war period, Rafsanjani was perhaps Iran's most popular statesman, but this account of his democratic credentials is incomplete. It neglects to say, for instance, that his only opponent in the 1989 election campaign had urged the people to vote Rafsanjani. It does not mention that, at the time of his election to the Assembly of Experts from the constituency of Tehran, the field was one more than the number of seats to be filled. His understanding of democracy did not privilege the liberal Islamists who, in 1991, criticized him in an open letter; twenty-three of them were arrested.

At the 1997 presidential elections, the conservatives didn't bother putting up a candidate that might be attractive to the people. Their man railed about Western 'arrogance' and 'deviations in the name of freedom'. The Guide contributed a tirade against the 'cultural and political attacks of foreigners'. It was all designed to put people off Khatami, who was portrayed as the foreigners' Trojan horse. But Iran had changed. Millions of first-time voters couldn't remember the Revolution or the Nest of Spies. Millions of women wanted equality and respect. These people voted for Khatami because he promised to make elected institutions answerable to the people. They voted in such numbers there was no chance of rigging.

Upon hearing the results, a friend of Mr Zarif's pounded his head against the wall and wailed, 'It's over! It's over!' For a while, it looked that way. Khatami explained on CNN that Iran's reformists sought what the Puritans had sought when they landed in America. (Freedom and God.) There was Montazeri, reminding the Guide that he wasn't an object of emulation and never would be, and attacking Iran's 'monarchical set-up'. (That was the last straw; Montazeri's office was ransacked and his seminaries closed, and he was placed under house arrest.) You had Mohajerani, who spoke well of the novels of Milan Kundera, whose daughter played the piano.

About that time, 1998, Mohsen Kadivar, the handsome theologian, brought out a book in which he examined the ten sayings, attributed to the Prophet and the Imams, which are commonly presented as evidence for the necessity of clerical rule. Kadivar concluded that eight of these sayings might not be authentic. Even the two that he considered 'authoritative' could not, he argued, be used to justify the clergy's assumption of political power. Rather, they conferred on jurists the responsibility to 'propagate, publicise, and teach Islamic rulings . . . in conclusion, there is no authoritative evidence for the absolute guardianship of the appointed jurist.

Something had to be done.

They killed Darioush Forouhar because he made no secret of his belief that religion should be separate from government. He'd declared a limited opposition to capital punishment, a declaration that some equated with apostasy. His political party publicized human rights abuses that were taking place in Iran. They killed him for his moustaches, unwaxed but rising at the tips. Like his ideas, they were a loose adaptation of a Western exemplar, an inadmissible rejection of their bearded, Islamic way.

Parvaneh Forouhar was an irreligious poet. She regarded the compulsory Islamic head covering with loathing. She would have raised hell to find her husband's killers. All good reasons for her to die.

There were four other victims at the end of 1998, all writers: Pouyandeh and Mokhtari (suffocated); Davani (disappeared); Sharif (suspicious heart attack, rumoured to have been induced). One of them wrote verses that, it was felt, concealed subversive thoughts. Another translated European texts about human rights. Some had been involved in the Pen Association, a suspicious guild.

They weren't reformists. Forouhar doubted that Khatami, a cleric whom the regime had nurtured, had the will or ability to change Iran. But he and the others had been the indirect beneficiaries of reform. In the name of strengthening the regime, Khatami and Mohajerani were allowing its opponents to breathe.

The killers had reckoned without Khatami's newspapers. They found their actions minutely examined and denounced. Journalists fell over each other in their enthusiasm to reach the truth. 'Find the killers!' they demanded. 'Who gave the orders? This is nothing short of an attempt by Khatami's enemies to hound the president from office!'

Khatami declared, 'When we don't accept someone, we make of him a counter-revolutionary, a monarchist, corrupt, pro-Western, a threat to national security and an apostate. Then, if some ignoramus says this counter-revolutionary must be killed – well, they kill him.'

The conservatives manufactured theories. The perpetrators of the Forouhars' murder were friends of the family; the killers were linked to Iran's exiled opposition, counter-revolutionary elements and foreign journalists; the writers had been killed by people who favoured reopening relations with America; the Turkish government had arranged for Forouhar to be killed to pay him back for his sponsorship of Kurdish separatism inside Turkey; the victims had been sacrificed by their paymasters, the CIA, in order to wreak havoc inside the Islamic Republic; Khatami's supporters had committed the murders to discredit the conservatives.

'Even in the furthest recesses of the brain,' *Kayhan* said, 'there can be no room for the suspicion that the perpetrators of Forouhar's murder were from the forces of the Revolution! In fact, there's no logic behind the idea that revolutionary forces could have been behind a counter-revolutionary action!' It was rumoured that the people at *Kayhan* were shredding a vast quantity of newspapers every day. It was a basic way of protecting a print run which had been computed according to machismo and nostalgia, not realistic sales expectations.

Khatami deputed three trusted men to investigate the murders and report to him. He was taking no chances with a conservative judiciary that might whitewash the business. On 4 January 1999, under pressure from the president, the Intelligence Ministry announced 'the involvement in this affair of a handful of irresponsible, evil-thinking, deviant and obstinate figures within the ministry. These irresponsible figures committed criminal acts as the agents of secret elements, and with the aim of fulfilling the designs of foreigners. The perpetrators of the murderers have not only betrayed the warriors of the twelfth Imam (may his return be hastened by God), but also struck a heavy blow against the prestige of the sacred regime of the Islamic Republic.'

The reformists demanded the resignation of Dorri-Najafabadi, the Intelligence Minister whom the Guide had foisted on Khatami. The minister went, albeit with exceeding ill grace. The reformists got bolder. Some asked whether the six presumed killings at the end of 1998 were the full extent of what were being called the 'serial murders'. What, they asked, about rumours of dozens of extra-judicial killings during Rafsanjani's presidency – murders that, they said, had been covered up?

This was the scenario that Akbar Ganji, a reformist journalist, presented to the readers of *This Morning*, the best selling of the progressive newspapers. Ganji named several people who, he said, had been murdered during Rafsanjani's tenure. Many readers hadn't heard of these people. What they had in common was a record of active or presumed political dissidence, and an abrupt end – an unsolved knifing or strangling, an unexpected heart attack, or disappearance. These deaths had either not been acknowledged as murders, or judicial investigations into them had been abandoned for lack of evidence. In some cases, the families of the deceased had been advised to drop their enquiries.

Gradually, as more and more journalists picked up the threads, Ganji's readers started to hope: that the truth about all political killings would be uncovered, that the men running the Islamic Republic would be made accountable.

There was a counter-trend. The case was transferred from the relatively transparent criminal courts to the military courts, which have a reputation for secrecy. Then, in the spring, Niyazi, the military prosecutor who had been put in charge of the case, made this announcement: 'In spite of the surveillance under which he was placed, Saeed Emami, one of the pivotal masterminds of the murders, committed suicide during bathing period on Saturday in the detention centre, by swallowing hair remover.'

Few Iranians had heard of Saeed Emami but they understood the meaning of his death. Suddenly, it seemed less likely that they would learn who had ordered the 1998 murders, let alone find out about those that had allegedly been committed before. Saeed Emami would take the information to his grave. One of the newspapers published his grainy photograph, taken when he was young. Through his glasses, you could see his pupils, shining darkly.

❖   ❖   ❖

There's an amateur film of a speech delivered by Saeed Emami in a hall in Hamedan, in western Iran, that took place on or around 23 December 1996.

The cameraman has missed the first part of the speech; when the film starts, Saeed is in full flow and the camera is panning across the audience. To the left of the hall, neatly segregated, are the ladies, almost all in chador. When the camera alights on the men, who make up the majority of the audience, you notice a cleric, sitting in his gown and a ruffled turban, coping with a winter cold. The rest of the audience are non-clerics; they wear thick coats and trousers, and different models of beard to show off their virtue. Their listless eyes attest to a revolutionary zeal that is being pounded by doubts. Having been perfectly dynamic during the 1980s, the Islamic Republic is now inert. No, things are worse than that; Iran is being disabled from within, by liberals intent on corrupting and Westernizing; they use phrases like human rights and representative democracy.

Now, in their hurt state, the old soldiers have crowded into a lecture hall to hear Saeed. They have in mind, perhaps, Khomeini's admiration for the Intelligence Ministry; he called its employees the 'anonymous foot-soldiers of the twelfth Imam'. It's true; at least the intelligence boys give the impression that they haven't abandoned the struggle against vice; they seem willing to go on the offensive, to strike pre-emptively against the Hypocrites (Khomeini's name for the Peoples' Mujahedin) and other enemies. 'Of course,' says Saeed, 'we killed Hypocrites, and we've killed members of other groups as well . . . what are we to do with the Hypocrites: sit down and play chess with them?' If anyone can provide the veterans with words of comfort, and remind them of the values of the Revolution, Saeed's the man.

Which is ironic, for Saeed wasn't in Iran for the Revolution. He was studying – in America of all places, at Oklahoma University – and he followed events on the networks. After graduating, Saeed went to work for what remained of Iran's representation in Washington, and then at the mission to the UN. Already, he was passing on information to the people who would later set up the Intelligence Ministry in Tehran. Saeed Hajjarian, one of the guiding lights behind the ministry, and the man who interviewed Saeed for this work, found him to be a 'clever person' – not deeply spiritual, but observant. He was careful, for example, not to eat food that had been cooked in lard. Hajjarian is said to have recommended that Saeed not be promoted to a responsible position.

But that's what happened, after the War. (Saeed didn't fight; he was tracking down enemies of the Revolution.) On Khamenei's suggestion, Rafsanjani chose Ali Fallahian, a cleric, to be his Intelligence Minister. It was a controversial choice; Fallahian had a reputation for brutality. He organized a shake-up at the Intelligence Ministry, and Hajjarian and his friends, who wanted the ministry's work to be transparent and its employees to be accountable, were sidelined. The new minister took to Saeed. Before his thirtieth birthday, Saeed had become one of Fallahian's deputy ministers.

It's possible that Saeed regretted not being at home for the Revolution, not witnessing the time that Iran and Islam were reborn. This regret, and the desire to distinguish himself in the eyes of those who had participated, may have spurred him to greater zeal.

In the Hamedan speech, he wears a beard and square-framed glasses high up his aquiline nose. When he cracks jokes, there's a flash of teeth and a lick of tongue across voluptuous lips. The way Saeed talks makes him at once one of the boys and way above them; his Persian is colloquial, because the Revolution was supposed to answer the call of God in a way that's intelligible to man. But he patronizes his audience, too, because he's risen fast and he has his pride. By the end of his hour and a half, he's conducted his listeners on a *tour d'horizon* of his prejudices, his fancies and his faith.

Saeed is in his late thirties, and no one can be appointed to a senior position in the ministry without his agreement. He dabbles in the arts; state television has run the 'confessions' of opponents of the regime, films made by Saeed. He's active in foreign affairs. In 1995, his lads carried out a jolly raid on a dinner party being held by the German cultural attaché. Later, there will be the case of Hofer, a German businessman caught with his pants down – a capital offence in the Islamic Republic. There will be the affair of the ten Jewish 'spies'; their arrest, trial and conviction will upset Iran's relations with most of the rest of the world. All this was among the results of the mindset created by Saeed.

For Saeed, the man opposed to his point of view is like misshapen clay. The strength in his fingers may derive in part from the close relationship he enjoys with his boss. It's possible that Saeed experiences a kind of joy when he breaks his subjects – once he's taken their honour and they're begging for death. But most of all, Saeed is driven by love. He loves the Revolution, and he loves the memory that he cherishes of Khomeini, a love he has transferred to Khamenei.

Saeed announces that he's had conversations with intelligence agents from foreign countries. He's sparred verbally with former agents of Savak. At the ministry, he says, they have seventy-six TV channels; they monitor filth and slander being peddled all around the world. The audience nods gravely. The state has a duty to know its enemies and only exceptional people can perform this task without being corrupted. The audience doesn't want proof that things are as Saeed says they are. They've been relieved of the part of the brain that wants explanations and receipts.

Saeed's authority grants him what many Iranians crave: a rapt audience as he rambles about subjects he barely understands. Saeed says he favours an open political system. (No one asks: 'And what place does the killing of Hypocrites have in such a system?') He gives a short appreciation of the films of Steven Spielberg, which he likes, despite Spielberg's being a Jew. (The people of Hamedan are impressed; few of them have seen a Steven Spielberg film.) In passing, he corrects misapprehensions that exist on the number of Jews killed by Hitler. The figures you hear, says Saeed, are put out by the Zionist entity to justify its existence. The true number of victims is a quarter of a million. The audience absorbs this fact.

Saeed's appreciation of his own worth is apparent from his conversational style of address, from the easy lie of his body in the chair on the low platform, from the audacity of his boasts and lies.

'I want to give you the example of Saeedi Sirjani. The Saeedi Sirjani that came into being after we were pals – and there's quite a story there!'

Everyone remembers Sirjani. He was a writer and degenerate. (*Kayhan* said so. It must be true!) *Kayhan* revealed that Sirjani worked for Savak before the Revolution. *Kayhan* called Sirjani, 'counter-revolutionary, trai-torous and perverted'.

There was a privately published book in which Sirjani inserted a very important cleric into the story of Sheikh Sanan, the diverted dervish. Everyone knows about Sheikh Sanan and his Christian girlfriend; in the noon of his obsession, she causes him to lose sight of his faith. In Sirjani's retelling, the important cleric was the dervish and the girl was political power.

It's impossible not to experience a surge of gratitude towards the men – Saeed's men – who induced in Sirjani his eleventh-hour conversion. They put Sirjani before the TV cameras and he apologized for the things he'd been saying and writing. He announced: 'The New Sirjani is not the same as the Old Sirjani!' It was little short of miraculous. There were letters to the newspapers and a full examination of his crimes and the extent of his penitence. By the time Sirjani had died in jail, he'd been saved. (It was typical of the foreign radio stations and local gossipmongers to make out that Sirjani had been knocked about to get him chatting, and that his heart attack had been induced.)

'When he first came to us,' Saeed says, 'after the boys had picked him up from the drugs combat HQ, he really thought the Intelligence Ministry would thrash him, that there'd be a set to ...; he said, "go on and beat me, cut me to ribbons, I'm not telling you a thing". The boys said, "Listen, this is the Intelligence Ministry, and we're not into beating here. Here's a pen and some paper; go and write."'

According to Saeed, Sirjani wrote a poem addressed to the Guide, asking why Iran's men of letters were being degraded. One of Saeed's boys replied with a poem of his own. When Sirjani read this reply, its declaration of religious and political faith induced in him a dramatic transformation.

'Ten minutes later,' says Saeed with a fond smile, 'he rapped on the door of his cell and said, "Take me out of here; I want to confess." I said, "Confess to what?" He said, "I want absolution."

'You know, we've got fifty hours of him on tape and video ... he wrote one thousand seven hundred pages on why he had waged war against the clerics. On why he'd been an enemy of the Revolution. Those images we got of Saeedi Sirjani ... sometimes I sit down and watch them, and I weep.'

Sirjani's penitence was so profound that he was moved to request a trip to the front. 'The boys took him off, to the section of the front where they were exhuming the corpses from the trenches ... and he sat on a pile of corpses and cried his eyes out ...

'At that time, he didn't know what my position was in the pecking order. I would go every now and then to the house we'd rented for him ... there was a pool and some flowerbeds. He'd get up in the morning and pick flowers, and water them and that sort of thing.

'One day he started to pour his heart out, and he said to me, "I'll tell you something; I'm really dying for some dates and ground sesame ..." The following morning, I went to south Tehran and traipsed around for God knows how long until I'd found some ground sesame, and brought back the ground sesame along with some dates. And afterwards, when he'd realized what my post was (not that my position is anything but humble – far be it for me to contest the role I have been allocated by the Islamic Republic!) and that I'd done him this service, he said, "I can't believe you went all the way down to the south of the city and traipsed around on account of some ground sesame and dates."

'And I said, "since I sensed that you had started to understand our values, I wanted to indulge you".'

Saeed and his men tortured Saeedi Sirjani to extract confessions for broadcast. Then, they killed him, because he was scum. As Ruhollah Hosseinian, Saeed's old pal from the ministry, put it: 'Saeed Emami really did believe that the enemies of the Islamic Republic should be put to the sword.'

❖ ❖ ❖

The event was a symposium organized by Armenia's Pen Association, in 1995, and they invited all the distinguished Iranian writers and journalists they could think of.

In the days leading up to the departure, lots of the writers and journalists pulled out – a suspicious number, it's now thought. One participant said his wife had had a car accident, and it turned out later that she hadn't. Others came up with flimsy excuses.

Twenty-one of them were not tipped off. That was the number of writers and journalists on board when the bus set off for the border, after stopping for a break in the Caspian town of Rasht.

The driver, Barati, was a short, dark-skinned fellow, and he had the livery lips of the habitual opium user. As they headed into the mountains, and evening turned into night, some of the writers felt like another break. Barati promised to stop at a roadside café where they could procure bread and milk and delicious local honey.

They cleared Astara early the following morning. Most of the writers were asleep. Some of them dozed and wondered when they would get their milk and honey.

Suddenly, the bus lurched to the right and came off the road. Those who were awake thought they had reached the roadside café that Barati had described. Then Fereshteh Sari screamed. She screamed because Barati had put the vehicle into neutral and leaped out. The bus was trundling towards the darkness.

Massoud Tufan, who had taken the two seats at the front on the right-hand side of the aisle, jumped to his feet and wrenched back the handbrake. The bus came to a halt, they established later, a few metres short of the cliff.

The only explanation for what happened next is confusion. Not all the passengers were awake. Some of them thought Barati had fallen asleep at the wheel, lost control and leaped out in a panic. They made no attempt to stop him when he climbed back into the bus, turned the engine on and reversed towards the road.

Then he stopped the bus and put it back into first gear. He put his foot down on the accelerator. Once again, as the bus trundled towards the cliff, he put the bus into neutral and leaped out.

This time, a rocky outcrop at the angle of the cliff arrested their progress, catching the underside of the bus and preventing it from going over the edge. The front wheels hung in the air.

The writers got out of the bus. A few of them tried to lynch Barati. The others pulled them off; killing him would compound their problems.

As they waited for the police in a nearby café, some of them saw a black Mercedes-Benz approach, slow down near the spot where the bus had almost gone off the cliff then speed off. The local police arrived and asked some questions. Later, Alikhani, Saeed's right-hand man, arrived with some of his boys and shooed the police away.

They took the writers off to a jail in Astara, and they took Barati away. Alikhani interrogated them in a cell. He asked:

'What makes you think the driver was trying to kill you?'

'What is the purpose of your visit to Armenia?'

'What is your evaluation of the current cultural climate in Iran?'

Then he let them go.

You've got to hand it to Saeed. He's a performer, a real professional. That's clear when you look at the dates. The time of the Hamedan speech,

December 1996, is the time of the Sarkouhi screw-up. It's a screw-up that will give Rafsanjani, who resents Saeed for sabotaging his foreign policy of tentative détente, a pretext to demand his demotion. But Saeed isn't breaking sweat!

Faraj Sarkouhi is a magazine editor. His wife and kids live in Hamburg. He keeps a mistress in Tehran – a lady called Parveen, who, like Sarkouhi, is a member of the Pen Association. Sarkouhi has had dealings with the ministry, mostly chats. Roughly fifty days before Saeed's Hamedan speech, Sarkouhi set out to visit his family in Hamburg.

And that, Saeed explains in Hamedan, is where the problems started. Although it was proved that Sarkouhi had handed in his departure slip and collected his foreign currency entitlement at Tehran's Mehrabad Airport, no one, least of all his wife, was able to confirm that he'd arrived in Hamburg.

For forty-five days, Saeed informs his audience, he and his boys did their best to track Sarkouhi down. Meanwhile, there was talk about his safety. Human rights organizations were concerned. It was an opportunity for enemies to do down the Islamic Republic.

A few days before the Hamedan speech, the mystery was cleared up. Sarkouhi turned up at Mehrabad and announced that he'd not been in Germany at all, but in Turkmenistan.

'And, all of a sudden,' gloats Saeed, 'Reuters, the Associated Press and human rights groups like Middle East Watch and Amnesty International, and various other organizations which had been discussing him, suddenly took back what they had said and went quiet.'

Here's another version of the same story:

At 10.30 p.m., on 2 November (the day before Sarkouhi was due to catch his flight to Hamburg), Alikhani rang Sarkouhi to inform him that he wanted to see him at the airport early the following morning, before the plane left.

A month and a half before, Alikhani had taken Sarkouhi in. Saeed Emami had come up close enough for Sarkouhi to smell the kebab on his breath, and whispered, 'We're going to sacrifice you, to encourage the others.'

Why had Alikhani called him now, a day before he was due to fly to Hamburg?

He got to the airport at four in the morning, and one of the boys took him to a room where Alikhani got him to fill out and sign an embarkation slip. Alikhani took the embarkation slip, as well as Sarkouhi's passport and his foreign currency entitlement form. They blindfolded him.

In another room, they removed the laminate from Sarkouhi's passport photo, peeled off the photograph and replaced it with one that belonged to someone else. Then they gave Sarkouhi's passport to the other man. He drew the foreign currency that had been allocated to Sarkouhi and took the flight to Hamburg.

By that time, Sarkouhi was in one of those bland houses around the place, whose secrets and perhaps even existence are revealed to few state agencies. Later, they played him a recording they'd taken of a telephone conversation between his wife in Hamburg and his brother in Tehran, after he failed to take the flight for Germany. Sarkouhi heard his brother say that the immigration authorities in Tehran had confirmed his departure.

A couple of days later, they told him, 'You've been recorded missing, but you've gone missing in Germany, because we can prove you left the country and that you caught the flight to Germany. We're going to keep you here while we chat to you and, after we've finished chatting, we're going to kill you and bury you where you won't be found.'

He signed everything they wanted him to sign, and he noticed that they dated the confessions to the previous occasion he'd been arrested – the time he'd met Saeed. They got him to talk about the Pen Association, and about his sexual relations – about Parveen – and he wrote down whatever they told him to.

While the cameras rolled, they got him to say that he'd spied for the Germans and that his wife was receiving a German government stipend on his behalf. They got him to say that the Pen Association was being directed by Germany and that its goal was to undermine the Islamic Republic.

Why?

Mykonos, of course! It was the name of the restaurant in Berlin where four Iranian Kurdish dissidents were killed in 1992. At the time of Sarkouhi's arrest, a German court was trying Iran's leaders *in absentia* for ordering the murders. (The following year, Kubsch, the judge, would find them guilty of a 'flagrant violation of international law'.) It was an intolerable affront to the Islamic Republic, and Saeed, who loathed Germans, was coming up with ways of avenging the slight.

According to the plan, Saeed's boys would get all they could from Sarkouhi, kill him and broadcast his confessions. They would tell viewers that the recording had been made between 10 and 12 September – the previous occasion that Sarkouhi had been detained. The public would be informed that the decision to allow Sarkouhi to flee had been an oversight. Sarkouhi's disappearance in Germany would be attributed to the shame he felt as a traitor and philanderer. Saeed and his boys would sit back while patriotic citizens denounced Sarkouhi's treachery. Iran's relations with Germany would deteriorate.

Something went wrong. German immigration didn't confirm that Sarkouhi had entered the country. Without that confirmation, Saeed and the boys couldn't kill him and broadcast his confessions. It was very irritating. Saeed's lads came up with the Turkmenistan tale and let him go.

After that, Sarkouhi was free but he felt imprisoned. He couldn't talk to the media. The local papers wouldn't dare print what he alleged. The foreign press was tainted. He sensed the ministry – listening, watching. It was only a matter of time before they tried again.

Sarkouhi wrote a letter, which began, 'I was arrested on 3 November, at Tehran's Mehrabad Airport, and I was imprisoned until 20 December, in a secret prison attached to the Ministry of Intelligence.' Then he described the events summarized above.

At the end, he wrote, 'If someone should come into possession of this document after my arrest, they should make sure it reaches my wife three days after my arrest, or the day after my death, so that she can have it published. If no one receives this, I am dead. In fact, I have been dead since 3 November 1996. I love my wife and my children dearly, and my life until 3 November was an honourable one.' He gave the letter to a friend. If he was arrested again, the friend was to ensure that the letter reached Hamburg.

Sure enough, they took him in and the letter got to Hamburg. The foreign radio stations read it out and Saeed realized that he'd been outwitted. The judiciary took Sarkouhi out of his hands. They sent him to jail for a year – his letter had contained 'propaganda against the Islamic Republic'. Then they freed him and he went to Germany.

In the summer of 1999, Saeed died, as many of his subjects had died – away from prying eyes, after interrogation at the hands of agents from the Intelligence Ministry. Then they shovelled dirt onto him. They said he'd been responsible for a bombing in Mashhad, that he and his wife had visited the illegal Zionist entity. They arrested Mrs Emami and forced her to confess to being a Mossad agent and a loose woman, and they forced her to admit that her husband had been a paedophile. They kept hold of these confessions, to dissuade her from going public with dirt of her own.

In the summer of 1999, before the conservatives lost control of parliament, the deputies discussed an amendment to the press law, whose ratification would make it easier to send journalists to jail. On 6 June, *Salaam* published a letter that had allegedly been written the previous October by Saeed Emami to the then Intelligence Minister, Dorri-Najafabadi. In the letter, Saeed suggested an amendment to the existing press law – the amendment's ratification would make it easier to send journalists to jail.

On 7 June *Salaam* was banned for printing a classified document. At Tehran University students gathered to protest. Conservative vigilantes and the police attacked them and they fought back. For a few days the unrest spread to other cities. Two people were killed and hundreds arrested. The Revolution seemed under threat.

❖ ❖ ❖

Akbar Ganji faces Saeed Emami. At the beginning of the Revolution, they might have got on. But Saeed and Ganji became opposites, or perhaps not opposites but definitely enemies, and it's the story of how the Revolution destroyed itself.

After 1979, Ganji joined the Revolutionary Guard; because he was an ideas man, they assigned him to Doctrine and Politics, which meant instilling revolutionary values into people. Later, he did a stint at the embassy in Ankara, supervising the translation of Persian classics into Turkish (so they say). After the War, he got friendly with Alavi Tabar and the crowd at the Centre for Strategic Research. (It was a stellar little group; it included Kadivar, the handsome theologian, and Hajjarian, Ganji's friend who had been sidelined from the Intelligence Ministry.) He started a magazine that promoted New Religious Thought, which was in vogue. New Religious Thought made much of the flexibility of Islamic law and its ability to evolve in reflection of a democratizing society.

When I came to live in Iran, Ganji became my hero. Two of his books, which were dominated by articles about political murders in the Islamic Republic, were the first books in Persian that I read. They were as close to him as I could get, since he was already in jail. I envied those of my journalistic colleagues who had met him before his arrest. He sounded slightly off the wall.

At the end of 2000 I saw him in the juryless Revolutionary Court, which tries cases deemed to have a bearing on national security. Ganji was being tried, along with several others, for taking part in a conference in Berlin that had been upstaged by counter-revolutionaries. No one was sure if he would turn up; at a previous hearing, in the Press Court, he'd refused to wear prison clothes and the judge hadn't let him in. Suddenly, there was the sound of feet being scuffed along the linoleum and some grunts. We turned around and saw a man inside the door being manhandled by two bewildered conscripts. He shouted: 'They tortured me! I've been tortured!'

Ganji was directing his comments at us, the foreign media. When we got back to our offices that afternoon, our editors rang to say that they'd seen newswire headlines flashing TORTURE and MAVERICK JOURNALIST. But it was pretty obvious that Ganji hadn't really been tortured – not in the vicious, premeditated sense that our editors had in mind. He probably got some fists in the face as the warders stuffed him into his prison uniform.

He enjoyed his day out. He harangued the judge and public prosecutor, whose roles seemed interchangeable. He was sharper and wittier than either of them. He refused to shut up even when the judge entreated him to. He grinned as the prosecutor read out a charge sheet featuring 'anti-national activities' and 'propagandizing against the regime'. He had his back to us but every now and then he'd turn around and tip up his chin, the way little men do when they want to see across a crowded room. He wanted to catch the eye of his wife and brother, and to see which ministry people were there. He nodded at his colleagues from the domestic press. At the end of the hearing, he announced that the state had tried to blackmail him into stopping his investigation into the serial murders. They'd warned him that he'd get fifteen years if he didn't shut his mouth. (And that's what he got a few weeks later: ten years in jail and five years' exile somewhere hot.) After the hearing, Ganji was carted back to the cells, his prison jacket tucked petulantly under his arm.

Between January 1999 and his arrest in March 2000, Ganji was the most controversial man in Iran. It wasn't only his repeated demands to know why the judicial investigation into the serial murders had been limited to four killings. It wasn't merely the light that he shed on some eighty murders that allegedly took place between 1989 and 1997. Ganji did more than reveal. He taught Iranians to demand of their rulers something they'd never dared to demand before: an explanation.

He visited the newspaper archives. He went through the anonymous letters he was receiving, deciding which were worth following up and which designed to mislead him. He spoke at length to Hajjarian, the former Intelligence Ministry official who was on the editorial board of *This Morning*; Hajjarian is said to have provided Ganji with some of his best material. Others on the editorial board, including Alavi Tabar, suggested that Ganji leave his articles unsigned, but he refused. His revelations were a sort of theatre and he enjoyed taking a bow. Besides, fame gave him a kind of safety. The greater his notoriety and the more blatantly his enemies were obliged to loathe him, the better were his chances of staying alive.

His colleagues revered him because he laughed at the threats. It buoyed them to know that he feared death no more than going to the dentist. They went out and gathered information for him. They were aware that this was the golden age of Iranian journalism, and that their product sold two hundred and ninety thousand copies a day – a figure, already astonishing, that would have been higher if only the presses worked faster. They were aware, too, that the golden age could come to an end at any moment. Iran could be silenced once more.

Ganji gave us Saeed Emami boasting in the company of important officials about the way his men had murdered Saeedi Sirjani. He gave us Saeed cooperating with *Kayhan*'s editor to interrogate subjects, and creating the famous television confessional. He gave us Saeed destroying his subjects, Saeed with interesting pseudonyms (Kooshan; Muhammad Ali; Kashigar; A Senior Bureaucrat), and Saeed openly advocating the murder of five or six of Khatami's close associates. He gave us Saeed planning the early serial murders and guiding the latter ones. (The killing of Forouhar and the four writers came after his demotion.)

None of this meant that Ganji could prove all, or even most, of his claims. It would have been crazy for him to reveal his sources. It may be that he had little, if any, documentary evidence for what he was writing. He put down what happened, quite baldly, and hoped that he would be believed.

One day in *This Morning* he told his readers about Fatemeh Ghaem-Maghami. Some people remembered the unexplained murder of this air stewardess and mother of three. How could she have been the victim of the ministry's scheme to murder political dissidents? It was true, Ganji said, that Ghaem-Maghami had no political past. But, she had been 'the witness of a regrettable event whose disclosure would cost the murderer and his ally dearly. So, there was no solution but to kill her.'

Ganji alluded to no document. His description of the way she died – shot, after a conversation with her killer that lasted an hour and a half – added nothing to previously published newspaper reports. But the Ghaem-Maghami piece was thrilling. The 'murderer' and his 'ally' knew who they were. Ganji knew who they were. Reading the piece was like eavesdropping for a few seconds on a private telephone conversation. There was Ganji, whispering down the line: 'I'm onto you . . .'

To help his readers remember who was who, he gave aliases to the villains and they found their way into Iran's imagination. Once upon a time, there was a Master Key, and he ordered the murder of people who used words in the wrong way – orders that Saeed and his men carried out. But the Master Key, Ganji said, had a second function. It was in his power to open the door to the Dungeon of Ghosts, which was the source of all the evil. Once inside, it would be possible to make out the Eminences Grises. They were influential people who used their public standing to denigrate a putative victim's character and mores, and to call into question his adherence to revolutionary ideals. In some cases, the Eminences Grises invoked Islamic law to pass unofficial death sentences on the 'accused'.

There was an Eminence Rouge. 'Louis XIV,' Ganji wrote, appointed Cardinal Richelieu to be his prime minister, and Richelieu gave Father

Joseph a senior position. 'On the orders of Father Joseph (the Eminence Grise), dissidents were murdered, and Cardinal Richelieu stayed silent. Because Cardinal Richelieu wore red, they called him the Eminence Rouge.'

Ganji managed his readers. He allowed the nimbler ones, who worked in government offices and picked up scraps of gossip, to guess the identity of the characters early on. That knowledge made them complicit in the revelation and keener still to find out how the story would end. For every measure of analysis and polemic, he gave them a half measure of thrill. It kept them on the edge of their seats and receptive to his wider sermon.

'Saeed Emami . . . took Siyamak Sanjari and a few others to a house, and he discussed matters there with him for a few hours. Then he ordered his men to stab him to death. Siyamak Sanjari started to cry and said he was due to be married in the next few days. Emami contacted the Master Key by mobile phone, and told him that Sanjari was weeping, and that his wedding ceremony was due to take place soon. "He even has invitations addressed to you and me in his pocket." The Master Key ordered Emami to kill him and they stabbed him fifteen times. Then they set fire to Sanjari's Mercedes-Benz in one of the valleys around Tehran.'

Put yourself in the shoes of the Master Key. To start with, you're affronted by your alias, which makes you out to be a mere facility. But, in the face of Ganji's strange and intimate knowledge, your irritation turns to apprehension. Other journalists pick up on your alias. Gossip is flowing through Tehran. What was going on between Ghaem-Magham and the Master Key? What unorthodox relationship meant that Sanjari had to die?

Ganji wants to see how you'll react, whether you'll be panicked into an indiscretion. He says he's not interested in retribution, that destroying the culture of violence is more important than destroying the perpetrators. On that score, you seem safe; Niyazi, the judge running the case, has made it clear that he'll look no higher than Saeed Emami and no further back than Darioush Forouhar. But Ganji could damage you. What if he has documentation to back up his claims?

Gradually, even those who hadn't already guessed came to suspect that, when Ganji talked about the Master Key, he was referring to Ali Fallahian, the Intelligence Minister. Ganji dribbled this into his readers' consciousness in a way that wouldn't get his newspaper banned. He used suggestive juxtapositions and allowed his hints to get broader. He got his public used to an allegation of astonishing impudence, without making that allegation explicit. (Until his final appearance in the Revolutionary Court, when he named Fallahian as Master Key.)

Fallahian flailed. He called Ganji a liar. He refused to condemn the killings. Then, one day, while explaining to journalists why he'd rejected an invitation from parliament to defend his ministerial record, he delivered a warning – to the man above him. He said: 'If it's decided that the minister should go and answer questions, he should go and answer them with his boss.'

Saeed had pointed the way to Fallahian. Fallahian showed Ganji into the Dungeon of Ghosts. There, Ganji found (and named) the Eminences Grises, all of them senior clerics and judges. Ganji didn't stop. He went to Rafsanjani.

The audacity! Rafsanjani wasn't just a former president; he remained one of the most powerful men in Iran and reputedly the richest. He was tipped to cruise back into parliament in the 2000 elections and become parliamentary speaker. Ganji had other ideas.

The 'Eminence Rouge', the first and most notorious of Ganji's Rafsanjani pieces, was not a very good article. It was uncharacteristically impatient; it tried to tarnish all of the former president's brightest buttons in one go. Ganji presented his evidence hurriedly, and a lot of the best stuff was submerged beneath the author's contempt. There was little, if anything, that was revelatory. The difference was that Ganji was addressing the whole nation, out of doors.

After the 'Eminence Rouge', Rafsanjani was no longer able to portray the Iran of his presidency as a just realm and interpret the nation's silence as a nod of agreement. No longer could he make out that he presided over (his words) 'the cleanest period in the Intelligence Ministry's history'.

Eighty-odd extra-judicial executions; the abductions and interrogations; Saeed's 'confessional'; the Armenian bus holiday; these took place, Ganji reminded his readers, during 'the cleanest period in the Intelligence Ministry's history'. Had the Rafsanjani years not coincided with Mykonos and – in the words of Ruhollah Hosseinian, Saeed's old work pal – 'hundreds of successful operations committed abroad by Saeed Emami'?

The Friday after the piece appeared, Rafsanjani accused Ganji of 'calling into question the prestige of the Revolution' – of nothing less than a 'vast, vast betrayal'. Rafsanjani's words did his reputation as much damage as Ganji had done. People wanted him to answer the allegations, not equate his own prestige with that of the nation. Even newspapers which had disapproved of Ganji's impudence were forced to be more dispassionate in their evaluation of Rafsanjani's tenure. They, and their readers, found it to be less glorious than they had led themselves to believe.

In the parliamentary elections, reformists swept the field. When the votes were counted, it was found that Rafsanjani had come thirty-first in Tehran, which has thirty seats. Even after Rafsanjani had been bounced up, by dint of a suspicious re-count, to thirtieth, he declined to take up his seat. It was Ganji's finest hour, the reformists' finest hour.

There was a price to pay.

A few weeks after the elections, a man riding a powerful bike of the kind used only by the security forces shot down Saeed Hajjarian and almost killed him.

Later that year, conservative judges banned about thirty reformist publications, including *This Morning*. They jailed more than a dozen journalists.

In January 2001, after a trial *in camera*, a judge sentenced to death the three Intelligence Ministry agents who had murdered the Forouhars and two of the writers. He sentenced two of Saeed's boys, including Alikhani, to life imprisonment. He jailed the opium-using driver, Barati. The families of the murder victims boycotted the trial, denouncing it as a whitewash.*

---

* They took the decision to boycott the trial partly on the strength of the perusal that Parastu Forouhar had made of the case file concerning her parents' murders. This perusal, in her own words, revealed the following:

Contrary to the assertions of some officials, the transfer of the case to the military prosecutor's office happened solely because of a written order issued by the head of the judiciary, and the case file contained no document concerning the legal basis of this decision.

Many pages pertaining to interrogations, and other documents, had been removed from the case file. They included the interrogations of Saeed Emami, who, at one time, had been described by the Tehran military prosecutor as the principal accused. Likewise, the clinical report on the cause of his death had not been appended to the file. In the confessions made by the other accused, the role of Saeed Emami in

Some of the reformists turned against Ganji. They criticized him for driving Rafsanjani into the arms of the conservatives, and for provoking the judges' offensive against the press. Ganji, they argued, had shown excessive disrespect to his seniors. He should have doffed his cap while he was calling them hypocrites, liars, murderers.

Many other people, however, continued to admire Ganji and to regard his imprisonment as a tragedy. They said Ganji represented the best and the bravest of their country. They said he had lots more information about the serial murders, and they expressed the hope that one day he would release it.

'Why don't you redecorate, change the look of the house?'

I was sitting with Parastu a few days before she was due to go back to Germany. She replied, 'The house has a contradiction in it. It's the place where two murders happened, but it's also somewhere that we try and make alive, to live and work in.'

---

these murders was left very ambiguous, but the interrogators asked no questions that might have shed light on it.

Any interrogation of Mostafa Kazemi and Mehrdad Alikhani, who were among the principal accused, about events that took place before 1998, had been removed from the case file.

... With reference to various witnesses and evidence, Mostafa Kazemi and Mehrdad Alikhani claimed that they had been ordered to commit the murders by Dorri-Najafabadi, the Intelligence Minister of the time. In order to demonstrate the organized nature of the murders, and their own subservient role, they cited examples of other similar crimes that had been carried out against dissidents, including the Armenian bus plot. No proper enquiry has been conducted on these affairs.

A number of the accused assert that the 'physical elimination' of dissidents was part of their job description, and that they had carried out similar actions prior to the autumn of 1998. In order to corroborate this, each and every one cited witnesses, but the judicial officials responsible for the file took no notice of these chilling confessions, and there was not a single question about these cases in the file ...

One of the accused asserted that, in his view, no murder had taken place, but rather the elimination, under orders, of 'two treacherous and dirty elements' (my mother and father). Because of the lack of sufficient evidence pointing to his active involvement in the murders, this defendant was acquitted.

Not too alive. She didn't give family dinners here. If she wanted to socialize, she went out. Friends popped in but they usually had some connection with the case or her parents.

Parastu waved her hand around the L-shaped room. 'Their presence in this room is a living one, but their death is also present, and their after-death.' There was another factor: justice that hadn't been done. Unless the real culprits were exposed, it was hard to imagine the house changing. It would be a monument to unfinished business.

The murders and the expectations that people had of Parastu – they'd created a disruptive energy. She would have to honour her parents. She would have to raise two adolescent sons. And her lover in Frankfurt – how much was he an observer, how much a participant? She'd need strength and cunning to satisfy them all.

'When you come to Iran, do you leave one Parastu Forouhar at the airport and become someone else?'

She shook her head while inhaling from her newly lit cigarette and waving the match through the air to extinguish it. 'It may have been like that at the beginning, but I've progressed. When I'm here, I have before me those things that are my life over there, and vice versa. I've tried not to allow my personality to fracture.'

A few weeks earlier, Parastu had tried to mount an exhibition, her first in Iran since the murders. There was a series of photographs of someone wearing an embroidered chador. But where there should have been the face of an Iranian woman there was a bald man's head – the back of his ugly head, pink and mottled with some stubble above the neck. In Berlin, they'd enlarged one of these photographs and put it on a hoarding by the side of the road: egg-like; obscene. The gallery in Berlin had larger photographs of the same person. He'd been sitting/kneeling with his head tipped back, which compressed the skin at the bottom of his skull so it became lips that smiled.

It was no surprise that, when it came to putting many of these irreverent images on show in Tehran, some men from the Intelligence Ministry deemed them at odds with the official designation of the chador as the most virtuous feminine garment. The exhibition was off, but Parastu's opening went ahead and the public found the walls of the gallery bare except for frames that had contained photographs of a bald head.

I asked Parastu if her father had been upset when she left for Germany. He'd been a nationalist, one who refused to flee in 1981, preferring perilous incarceration. On the contrary, Parastu said; her father had approved of the decision. Perhaps he realized that his daughter would be happier else-where. And it wasn't as if she'd stopped being Iranian. (Her children may have, though.) In fact, her vision of Iran may have become more coherent, her expression of that vision more lucid.

After a while in Germany she felt ready to start painting again. She found herself drawn to heavy colours and religious symbols. They were the motifs of the Islamic Republic, the state from which she'd fled. Now they'd been transplanted. They meant something else.

Later on, she and some friends turned a condemned building into a house of art. Onto a wall she painted an Iranian lady with Timurid eyes and bright damaged colours. The lady held a *daf*, a heavy tambourine, practitioners of which strike a seemly aristocratic pose, while irises wound their way up walls and across the ceiling. On another wall another lady stood on her hands.

On another occasion, she flooded floors and walls and a stairwell with a choppy sea of Persian calligraphy. When she showed me the photographs I thought of the fragility of words in a totalitarian state; visitors to the installation would have to tread words underfoot.

She laughed when I asked her if it was about censorship. I'd been in Iran too long! What mattered was the visual rhythm of the words. Parastu's relationship with her mother tongue was changing. Persian was no longer a language of everyday communication. (That was German, the language of her family, her thoughts and, increasingly, her dreams.) Persian had become a *hess*.

In 2002 she created a wicked little picture book.

It was revenge for being told: 'The answer to your question can't be divulged for reasons of national security.' It was revenge for having to remove her shoes and put on slippers and a smelly, borrowed chador when she entered a judicial building, and for the obfuscation and contempt she met inside. It was to pay back the judge who informed her that, if she invoked her Islamic right and demanded the death penalty for her mother's murderer, she would have to pay blood money to his family. (Iranian law has set a man's blood money at twice the level of a woman's.)

It was to get back at them for making her wait while the venerable judge busied himself with lunching, napping and praying; she should come back tomorrow, only tomorrow was the birthday of the seventh Imam, Musa al-Kazem (*salaam* be upon him), and thus a holiday and the following day was Thursday so would she please come back on Saturday after the long weekend when his excellency might – just might – see her for a few minutes?

The book is called *Take Off Your Shoes*, and its spare cartoons are often captioned with the orders that you receive when you try and get simple information from a building that belongs to the judiciary. 'Go to the security gate.' 'Hold up your arms.' 'Put on a chador.' 'Take off your shoes.' 'Sit up straight.' The book relates the (inconclusive) efforts of two women (Parastu and her lawyer) to penetrate the fog. They wait, are frisked vigorously, kill time in cafés while being observed by reedy little spies, and wait some more. They beg audiences with judges. Eventually they are allowed to peruse some files. They get nowhere.

The faces of the women in the little book are white blanks delineated by the outline of their chadors. Women standing or sitting together look like a school of penguins. The male clerks in their Basij uniforms are featureless save for peppery stubble that obscures their necks and faces. They stop lounging only when a judge sweeps past in his cleric's robes; they contrive a shambolic salute. (Why do they salute the judges? I don't know.) The judge, distinguished by his corpulence, rosary and an untidy splodge in the middle of his face (his beard), orders no interruptions today. For every page of disappointment and indignity there are three pages of waiting for disappointment and indignity.

'Could you ever come back to Iran? To live, I mean.'

'Not yet. Things have to change before I feel this is my home. When I come to Iran, I'm reminded of Fridays when you go to the cemetery and stand by the gravestone of someone who died.'

And yet she sensed incipient change – in the student protests. Normal people, she observed, were more openly critical of the regime than they had been at any time since the Revolution. Parastu wasn't the first expatriate

Iranian I'd heard make such observations. People living inside the country tended to notice the aspects of life that stayed the same.

She told me that the young man I'd seen leaving her house a few evenings before was involved with the student movement. He'd asked her to address a meeting that was being planned. He was waiting for her answer and she hadn't made up her mind.

I found myself counselling against. Parastu would lose the effective immunity that she currently enjoyed. (They were unlikely to arrest her as long as she confined her campaign to the issue of her parents' murders.) She would be associated with reformists who were trying to control and manipulate the student movement. She would be a political tool.

'Is your goal something beyond the exposure and punishment of the people who ordered your parents' murder?'

'Of course.'

It hadn't always been like that but she'd discovered that you can't fight one injustice in isolation from the others. Then I realized that there was a second reason why I'd advised her not to address the meeting. I had an idealized portrait of Parastu. Her courage had neither material nor professional object. Politics might stain it.

One of the vivid memories that I'd retained from my first interview with her, two years before, had been her reaction to my questions about Ganji. No, she hadn't read the books. (I had them in my briefcase, the juiciest bits underlined in pencil.) No, she hadn't met the man. I'd been puzzled by this indifference. Apart from Parastu, Ganji had done more than anyone to make sure that people still talked about the Forouhars. Later, I realized that she distrusted him. He and the reformists were using her father's memory to advance ideas that were not his own.

Since then, Ganji's bravery had made her revise her opinion of him. But her attitude towards the reformists in general had hardened. She'd had contact with reformist parliamentarians; they affected great interest in the serial murders and did nothing. The worst thing about these people, in Parastu's eyes, was their propensity to compromise. 'Sometimes it seems they're so concerned with their little deals, they lose sight of what they believe.'

She named a prominent reformist writer. He'd been to see her in

Germany. Parastu curled her lip. 'He started – oh! How he started! "Your father, the equal of the *seyyed*s and martyrs . . . blah blah . . ." The man had gone on and on praising her for fighting against injustice. Then, he'd said, "What about your poor lawyer, Zarafshan, languishing in jail? What can be done for him?"

'I told him, "It goes without saying that we'd do anything to help him. The reason he's in jail is the work that he did on the case. We've written to the UN Human Rights Commission and to the EU, and explained the situation. But, apart from that, what can be done?" He said, "No, madam! That's not what I mean. You sit down and do a deal with these people, and he comes out."

'I said to him, "We're not deal makers."'

CHAPTER NINE
Friends

One day Alavi Tabar invited me to come and see him at a management and planning college, at the northern end of the capital, where he taught in the mornings. The complex of buildings was grey and austerely neo-classical; it dated, I guessed, from the 1930s. The air was cleaner here than in the middle of the city. There were old oaks and shiny green verges around the campus, and a sense of studious calm. Here, Alavi Tabar wasn't a controversial newspaperman. He was a respectable teacher to whom the state entrusted young minds.

Sitting outside Alavi Tabar's office, I remembered the time when we'd first met, three years before. Back then it had seemed that the reform movement might work. It was possible that President Khatami, who had been elected on a pledge to make Iran more democratic, would succeed in steering a course between God and freedom, to the detriment of neither. No longer. Reform had been defeated. The conservatives had won.

The conservatives had feared that the reformists intended to take away not only their power but also their prestige, money and (perhaps) their freedom. Encouraged by Khamenei, Ayatollah Khomeini's successor as Guide, the Council of Guardians, the appointed upper house of hard-line conservatives, had vetoed every enlightened bill that parliament ratified. The judges had harassed, summoned and jailed dozens of reform-minded politicians, journalists and students; they had frightened thousands more into silence. At the beginning of 2004, they would go on to disqualify more than 2,000 reformist candidates, including scores of serving deputies, from standing in parliamentary elections. Conservative candidates would win handsomely on a turn out that was embarassingly low in the big cities.) It didn't necessarily mean that Khatami's 'religious democracy' was conceptually flawed. The conservatives hadn't allowed that to be determined, either way.

A few days before, I'd stopped by the office of an old friend, a canny car maker and economist called Saeed Leylaz. He blamed the reformists for underestimating the conservatism of Iranian society. 'They think north Tehran represents the whole of the country. Look at me, modern in my politics and economics; but if you asked me how modern I am in the way I live, I wouldn't know what answer to give.' I was reminded of what Mr Zarif had told me about a conflict, between tradition and modernity, which was taking place in all Iranians.

For the people, perhaps, democracy was a confusing idea. To appease them, Leylaz said, the state needed only to give them jobs. The younger generation was only slightly harder to please; they wanted jobs and the freedom to hold the hand of someone of the opposite sex in the street. Democracy sounded nice but many Iranians suspected that it might be used against them. How many Iranian men, for instance, are willing to share power with their wives?

I'd wanted the reformists to succeed. I'd wanted them to defy the Iranian exiles, sitting in LA, who summoned the people to rebellion through the medium of US-funded television broadcasts. I'd wanted them to disappoint America's neo-conservatives who, from a position of near-complete ignorance, wrote fluid little Utopias about a Middle East built anew in the image of New England. Later, when I became friends with disappointed revolutionaries, I had hoped that the Islamic Republic would evolve in a way that didn't humiliate them. I had willed the preservation of Iran's sole perceptible gain of the past quarter of a century: the liberty to take important decisions without having to consult a superpower.

The reformists were to have been the agents of this evolution and now they were unloved. People said that Khatami didn't have the courage of his convictions, that he was little different from the conservatives. His few achievements as president were the consequence of closed-door negotiation with Khamenei, a far cry from the transparency that he and his supporters were supposedly promoting. They said he was a good man, an intelligent man. But not a brave man.

As long as oil prices stayed high, Leylaz had told me, Iran's booming internal market would keep young people off the streets. But Iranians no longer felt that they had a say in the decisions that were being taken in their name; with the sense of powerlessness came a drop in self-esteem and civic pride, an increase in dishonesty, vulgarity and bitterness. You found it in wizened young addicts, in thirteen-year-old girls touting for custom at the side of the road, in soaring divorce figures and declining voter turnouts. There was nothing for it but to wait for the kings and princes to fall asleep on their thrones or be elbowed aside by other kings and princes.

Bearded goons broke up a wedding party; the bride and groom sat and wept. Rumours went around that university exam questions had been revealed to rich candidates. And the hypocrisy! The hypocrisy of denouncing the Americans from the pulpit while trying to do deals with them behind the people's backs; the hypocrisy of the judges who prevented women from divorcing their abusive husbands; the hypocrisy of sinking billions of dollars into nuclear facilities while the parks of south Tehran teemed with junkies whose welfare no government department was willing to underwrite.

This enfeebled Iran, said America and Israel, was a threat. You rarely came across an American newspaper article about the Middle East that didn't allude to the Iranian missile that was capable of hitting Israel. They made out that Iran's interest in nuclear weapons was offensive in design. (In fact, it was a card to use if America consented to play, an insurance against regime change.) They didn't acknowledge that the chauvinism propounded in Friday sermons was being betrayed in learned assemblies, convened to formulate policy, that were composed of the same sermonisers. The regime bellowed to belie its fading powers, while longing to bask in the sun.

I remembered, more than a year before, Alavi Tabar's prediction that he would soon be arrested. He had a suitcase ready; he'd packed everything he thought he'd need. (Every couple of weeks he would wash the clothes to make sure they didn't get smelly.) But Alavi Tabar hadn't gone to jail. Since the newspapers he was associated with tended to get banned, he stayed away from the press, writing mainly for reformist websites. He was teaching.

Some of his reformist colleagues, dynamic and combative, had been imprisoned. Often, they had come out different men, men who would henceforth have little to do with politics. I'd recently met a student activist who steered away from subversion, for fear of being imprisoned and 'burned out'. He would get his doctorate and a teaching job; that way, he told me, he'd be able to influence more people.

Sitting in Alavi Tabar's office, I found myself disparaging the reformists' commitment to their cause and their willingness to suffer for it. Alavi Tabar agreed; you got fewer heroes these days. (Akbar Ganji, journalistic investigator of the serial murders of the 1990s, had been the last.) Before the Revolution, young people had been imbued with a collectivism learned from the Communists: if you went to jail, you vowed, in the name of your comrades, not to break. Now you got liberal-minded reformists behind bars, wondering: 'Why should I carry the can?'

I asked him about torture. The Council of Guardians had spiked the government's decision, endorsed by parliament, to sign up to a UN convention against torture. An Iranian–Canadian journalist had been beaten to death while in the care of the chief prosecutor. But Alavi Tabar said that a lot of the stories you heard about torture were exaggerated.

'You can't compare the physical torture now with the torture before the Revolution. After the Revolution, you might get bastinadoed and roughed up, but that's about it. During the Shah's time, they'd pull out nails, attach electrodes to sensitive parts of the body ... what they called the Israeli methods.'

The Islamic Republic favoured a different torment. They disseminated information – true; false; half-true; what did it matter? – about sexual or financial improprieties. It was a way of destroying your prestige, your family. Take Mohajerani, Khatami's former culture minister. They'd spread rumours about his womanizing and driven him out of politics. Alavi Tabar told me of a forthright reformist deputy who, having been tarred with false allegations of corruption, found that his own father no longer trusted him. The deputy had become less forthright.

'What about Abdi?'

Opinion pollster, columnist and architect of reform, Abbas Abdi was known as the US embassy hostage taker who had repented. He had helped Akbar Ganji destroy Rafsanjani's reputation during the 2000 parliamentary election campaign. He'd called on reformists to quit public life. His polling company had made the discovery, unpalatable to conservatives, that 74 per cent of Tehranis favoured official dialogue with the US. When Abdi was charged with espionage in 2002, people expected him to turn the witness stand into a platform for defiance.

But Abdi's trial turned out like one of Saeed Emami's TV confessionals. He paid tribute to the governor of the notorious prison where he was being held in solitary confinement. He promised to 'atone' for organizing opinion polls that might have harmed the national interest. He retracted his earlier call for mass resignations. The reformists went into shock.

It was now well known, Alavi Tabar told me, that they had forced Abdi to recant by threatening to arrest his wife and daughter. Perhaps Abdi had feared that they would treat his wife the way they had treated Saeed Emami's. But that, in Alavi Tabar's eyes, didn't make Abdi's surrender excusable. Abdi had damaged his friends and colleagues – and the movement as a whole. 'If he was going to cave in, he shouldn't have spoken as he did.'

'And your family?'

I was asking whether he, too, would cave in. He replied, 'Abdi's wife is apolitical, whereas my wife sustains my politics. I often ask her, "What will you do if they take me? Don't think that people will come up to you and say what a good man I am."' It was always possible that they'd try and drag his name through the mud, too.

'How does she respond?'

'She's as ready to go to jail as I am,' he said proudly. I was reminded of the description he'd given of being stranded in a trench with six dead Iraqi soldiers. Each evening he'd looked up when the sun was setting and thought of his wife, knowing that she was doing the same.

Religion was Alavi Tabar's other sustenance. It would keep him going even after reform – after reform had become something else, perhaps more radical. I'd never seen his confidence waver in ultimate victory. Iran was a young country whose citizens no longer wanted what the establishment ordered them to want. It was a matter of time.

I was at home with our four-month-old son Jahan. I was having trouble connecting to my internet server; I had an e-mail to send to California. I carried Jahan to the kitchen and plugged in his bottle warmer. As I carried him back to my study he started bawling. I put him on my knee and phoned the server. They told me my credit had expired. I'd have to put money into their account and send the stub by fax. I put down the receiver and swore; I wouldn't be able to do that by the end of the LA office day. Jahan was yelling fit to burst, his fury almost propelling him off my knee, when the phone rang. Mrs Zarif said, 'I'm sorry for disturbing you, but I thought you should know that Mr Zarif's mother passed away.'

She'd been fading for months, after the doctors botched an operation on a tumour. I felt a sense of relief; Mr Zarif had been withdrawn for much of that time, in pain. In Iran, when the parent of a friend dies, it's customary, even if you didn't know the deceased, to turn up to at least one of the three memorial services that take place during the forty days after death. I asked Mrs Zarif about the first of these memorial services and she gave me the name of a mosque in east Tehran. I put the phone down and went off with Jahan to the kitchen.

Since Jahan's birth the Zarifs had been to our house several times; they'd gravitated to our newborn, perhaps to counter the death that they knew was fast approaching in their own family. Mr Zarif had told Jahan not to listen to his dad, who was filling his head with English words, but to his mum's Persian. I deplored the circumcision that had been pretty much forced on Jahan a few hours after his birth. The Zarifs were shocked when Bita said she was trying to get Jahan to abide by a set bedtime. In most Iranian households there is no bedtime, especially when there's a newborn babe to assault with pinches and cooing.

When Bita heard about the death of Mr Zarif's mother she asked if I had raised the question of our attending the memorial service or if Mrs Zarif had suggested it. Mr Zarif came from a bureaucratic family of quite high standing – part of the revolutionary establishment you might say. The members of such families do not have British friends; or if they have, they do not show them off. In the two years that I'd known him, Mr Zarif had never introduced me to his mother and father. I would stick out at the memorial service. What if Mrs Zarif had phoned merely to pass on the news of her mother-in-law's death, not expecting me to propose that we come to the service?

Three days later we went off to western Tehran. The mosque was flashy and new. Bita took Jahan and went to the women's entrance. I went to the men's side. Mr Zarif was first in a line of men waiting to greet people. I guessed that the old man next to him was his father. Seeing them, I felt a recurrence of my earlier worry: what if I wasn't welcome? Mr Zarif gave no clues. He was grey and bereft. He thanked me in a mechanical way for coming and I hurried in.

It was a subterranean hall that had been especially designed for such events. They're different from prayer halls in that you don't take off your shoes, and you sit on chairs rather than on the floor. The seats, arranged in rows facing a dais, were filling up. I went to the far end. I was fearful on account of my fair hair and blue eyes, but I told myself that my fear was irrational. I knew from experience that Iranians tend to see a foreigner only in places where they expect to see one. This was not such a place. I looked warily around.

The older men were standard revolutionary types. Their sons were more colourfully dressed and some were clean-shaven. I accepted a cup of orange juice from a man holding a tray. There was a mullah on the dais, speaking pieties into a microphone. He was besieged by lilies attached to wooden frames with sashes of the kind I have only seen on beauty queens. The sashes identified the institution that had sent the flowers. The mullah started reading the names of important people, followed by the message of condolence that they'd sent. There was a hierarchy, with university rectors and ambassadors to the fore.

Apart from a minority of close relations the men in this room hadn't known Mr Zarif's mother – it wouldn't have been proper. Perhaps they'd seen her. More likely, they knew of her as a safely accounted for appendage to a respectable family. Their purpose in attending the memorial service was not to remember her – what was there to remember? – but to express solidarity with the people who were remembering. So they did not weep or look put out that she was dead.

Ali, Mr Zarif's son, appeared at my side and whispered: 'My father wants to know if there's anything we can do for you.' Not understanding what he meant I shook my head and he sauntered off. Suddenly I felt perplexed and uneasy. Ali's choice of words had been odd – you might interpret them as hostile, an invitation to leave. But I could hardly walk out now. People were still arriving.

I cursed the Persians. Two years after I had first met him I still wasn't sure where I stood in relation to Mr Zarif.

A couple of rows in front two men rose from their places. They looked around, nodded unctuously at acquaintances and went out. I leaped to my feet and started walking along my row towards the door, feeling the combined stare of everyone in the room. There was no sign of Mr Zarif. I fled into the hot evening.

A few minutes later, Bita brought Jahan out. Bita asked, 'What's wrong?'

I told Bita what Ali had said and asked her what she thought he'd meant. She pondered for a second, and replied, 'Ali didn't mean to be hostile. It's polite to ask people who aren't part of the family or immediate circle if they're all right, if they need anything. Mrs Zarif did the same for me in the women's section. It's Ali's fault for putting it across in a sloppy way.'

'Are you sure?'

She smiled. 'Of course. Mr Zarif is your friend. Our friend.'

Hassan Abdolrahman had arranged to meet me at a quarter to six under the bridge at Karaj. The clocks had just gone back, so it was light – light enough to distinguish a floating film of soot and carbon monoxide from the pale sky, light enough to make out the leaden faces of men catching buses and shared taxis bound for Tehran, thirty kilometres to the east. I'd got to our meeting place fifteen minutes early, so I sat in the car under the bridge with the hazard lights on, fearing for my health as the buses and lorries pounded past and my tinny Kia Pride swayed in the foul wind.

I first clapped eyes on Hassan in *Road to Kandahar*, a film that was made in 2000 by Mohsen Makhmalbof and set in the Taliban's Afghanistan. Beyond confirming that Makhmalbof's creative inspiration had faded in proportion to his anger, this insipid film is distinguished by a single scene of riveting artistry, in which the female protagonist is examined by a male doctor, played by Hassan. The doctor is a black American Muslim who came to Afghanistan to fight the Soviets and never went back. This being the Taliban's Afghanistan, the examination must take place through a tiny hole in a curtain that has been drawn between doctor and patient. Through the hole, the doctor can do no more than inspect the woman's eyes and shine a torch into her mouth and ears. Their eyes meet.

After seeing *Road to Kandahar* I had tried to find Hassan. We had a mutual friend but Hassan was elusive. He lived in Karaj and came irregularly to Tehran. His telephone number changed a couple of times over the next year or so. When we eventually met, over lunch at the house of the mutual friend, I found him engagingly humorous and sardonic. We were both Western men who had married Iranian women and were living in Iran – I hadn't met anyone else who answered that description. We agreed to meet again, when I hoped to get him to talk about himself.

A few weeks later, when I was at Hassan's house for dinner, I proposed to him that we visit the citadel at Alamut. Hassan was surprised that I hadn't already done so; Alamut is one of the most famous sites in Iran.

In 1090, Alamut's seeming impregnability led Hassan-i-Sabah, who commanded a sect of heterodox Shi'as, the Nizari Ismailis, to make it his base for a campaign against the Sunni Seljuk Empire that was running Persia. Hassan-i-Sabah's aim was to establish what he considered the true Muslim Imamate under the leadership of the descendants of Ismail, the disinherited elder son of the sixth Shi'a Imam. The weapons deployed by his adherents included secret conversion, the concealment of their beliefs and political murder. So effective was this final tool that a later branch of Hassan-i-Sabah's order, the Syrian *Heyssessini*, gave a corrupted version of their name, 'assassin', to several languages.

The Nizari Ismailis' first and most celebrated victim was the great Seljuk vizier, Nizam-ul-Mulk, who was knifed to death by a zealot disguised as a mystic. In 1192, a Syrian branch of the order drew European Crusader blood with the murder of Conrad of Montferrat, king of the Latin kingdom of Jerusalem. As news of the Nizaris' murderous tactics spread, enemy princes and ministers felt increasingly vulnerable. Some wore chain mail under their clothes; it was not uncommon for Hassan-i-Sabah's agents to spend many years working for their enemies, apparently loyally, before receiving the order to act. Death would come from a dagger blow. Believing that their deeds earned them an automatic place in paradise, the assassins would make no attempt to flee, and this indifference to death heightened their psychological effectiveness. In the words of one Ismaili poet, 'by one single warrior on foot a king may be stricken with terror, though he own more than a hundred thousand horsemen'.

The passenger door opened at five forty-five. Hassan said, 'Salaam'. The Pride sunk a little when he got in; he'd put on weight since I'd last seen him. He told me that he'd been under a lot of pressure. It was harder to endure if you weren't in good physical condition. He moved the passenger seat all the way back, to make room for his long legs. I didn't ask what kind of pressure he'd been under.

Hassan Abdolrahman is another name for Daoud Salahuddin; he started life as David Belfield. When she speaks to him on the telephone, his mother in Long Island calls him Teddy, which comes from his second Christian name, Theodore. A quarter of a century since they last saw each other, Mrs Belfield might have trouble recognizing her second son now. His short wire-brush hair is greying and he's less physically impressive than he was as recently as 1996, when the *Washington Post* commissioned some photographs of him. They show a handsome man in a white polo neck under a double-breasted pinstripe suit, and it's hard to imagine him looking as sharp today. But these photographs, despite the ensuing physical changes, are as truthful as the day they were taken. They depict a man who has been detached or has detached himself. He stands among talking women, or among some people who are hurrying through a Tehran market, like an invisible watcher, a spirit that's being ignored.

Hassan is occasionally reminded of his notoriety. When *Kandahar*, as the film was renamed for Western audiences, got its US release, an Iranian–American who saw it recognized the black doctor as the same American Muslim who was wanted for the murder of an Iranian counter-revolutionary, Ali Akbar Tabatabai, in Maryland in 1980. Many people were offended that cinemas were showing a film that featured, in the words of the *New Yorker*, an 'American terrorist'. In the subsequent furore, one of Hassan's five brothers, who was working for the Internal Revenue Service in Montgomery, Alabama, received a visit from the FBI.

As we drove westwards, away from Tehran, Hassan was by turns laconic and loquacious. He had a way of starting rhetorical paragraphs with a stuttering repetition of the first word – 'I', for example, or 'the' – like the same note jabbing over a phrase of music, before starting to extemporize fluidly. The unobtrusiveness of his accent might be down to his long absence from America – or to the mutually neutralizing inflexions to which he was exposed: North Carolina (where his parents came from); Long Island (where he grew up); Washington, DC (where he lived for a decade until his flight).

I had no more difficulty imagining Hassan whacking a fellow named Tabatabai than I had had imagining Mr Zarif cracking heads with his *nanchiko*. During the three years that I'd spent in Iran, I had met many people who had committed acts of violence for their beliefs, and it no longer seemed shocking or distant. I knew nothing of Tabatabai, except for what Hassan had told the *Washington Post* and the *New Yorker* in interviews that he had given years after his flight to Iran.

Tabatabai had been a spokesman in the US for the Shah's final prime minister, Shapour Bakhtiar, a man who had tried to delay Khomeini's return to Iran and prevent the establishment of an Islamic Republic. Later, after fleeing to Paris, Bakhtiar had been implicated in a coup attempt, supported by the Americans, which had been aimed at overthrowing Khomeini. That attempt, led by air force officers, had been foiled, but it showed that many people, in and out of the country, would stop at nothing to topple the new regime. In Hassan's eyes, Tabatabai's links to Bakhtiar implicated him in this coup attempt – they made him an enemy combatant, someone who was fighting against Islam. There was virtue in ridding the world of such people. A few days after the coup attempt, disguised as a postman Hassan emptied his pistol into Tabatabai's stomach.

'Why,' I asked, 'would a well-brought-up kid with Baptist parents do a thing like that?'

The answer had less to do with post-revolutionary Iran than the experience of being black and growing up in America in the 1960s. This had made Hassan cynical and bleak. 'You know, back when Kennedy was smoking reefers in the White House, if you were a black man and you got taken in for marijuana, they came down so hard on you, you'd wish you'd been caught for killing another black.' Hassan felt 'separate and not quite equal', but was repelled by the demagoguery and indiscipline that afflicted militant groups like the Black Panthers and the Nation of Islam. As we drove, he described for me the self-inflicted fates of the three most prominent Panthers: 'Huey P. Newton: shot dead in the early hours of the morning while doing a narcotics deal. Bobby Seal: alcoholic. Eldridge Cleaver: died prematurely, lifestyle-related.' When he was very young, Hassan witnessed violence and social degradation at a nightclub, frequented by blacks, where his father worked. As soon as he was old enough, he started looking for alternatives.

In 1969, Hassan converted to Islam and took the name Daoud Salahuddin. In contrast with the blacks-only Nation of Islam, he found that genuine Islam didn't distinguish between the races and that it enjoined discipline and morality, as well as fervour. Six years later, his education was completed when he became the protégé of a brilliant Egyptian lawyer and militant who'd fought against the Jews to prevent the creation of Israel in 1948 and married the daughter of the founder of the Muslim Brotherhood. The man was short and immensely strong and his name was Said Ramadan.

It was Ramadan who came to know Hassan best – who was 'more alive, in more ways, than anyone I've ever known'. It was Ramadan who would, on the strength of a five-minute telephone conversation with Hassan's fiancée of the time, predict that she would betray him. (She did.) Ramadan's Islam didn't involve veneration for the Shi'a Imams but, in 1979, he told Hassan that the Iranian Revolution was important beyond Shi'ism; it crystallized the militant and religious universalism on which modern Islam would be built. He sanctioned Hassan's decision to become a security guard at the Iranian Embassy, at a time when counter-revolutionaries were rumoured to be planning to seize it. After Carter let the dying Shah into the US, Hassan joined five Iranians in chaining himself to the Statue of Liberty and unfurling a banner that bore anti-Shah slogans. (It turned out to be a lot of trouble for nothing: the protest coincided with a more newsworthy event, the takeover of the American embassy in Tehran.) After Tabatabai's murder, it was Ramadan's telephone call to Ahmad Khomeini that got Hassan his Iranian visa.

Now, almost a quarter of a century later, Hassan's anger towards the US still burned brightly; George W. Bush's adventures in Afghanistan and Iraq had seen to that. And yet, so much had changed. America was even more powerful and assertive than it had been in 1980, but Hassan seemed to have become equally disillusioned with Iran. In the past, I could imagine him militating in favour of the Iranians' right to develop the nuclear programme, allegedly military in intent, which threatened to bring them into confrontation with the West. But Hassan had become an observer and there was something knowing about the way he observed. It was no longer worth raising his voice in the Iranians' defence.

He had carried out his hit in 1980 on the understanding that the Iranians would send him to China to study two subjects that greatly interested him, Chinese medicine and martial arts. They hadn't done so. When Hassan arrived in Iran, they put him in a safe house. Later, he was allowed to work as a journalist, which gave him opportunities to go abroad. He travelled to other Muslim countries as Ramadan's envoy. Going abroad was a risk; it was always possible that the US authorities would track him down and spring a trap on him. He always had to come back to Iran.

And Iran turned out to be, in one way at least, unsettlingly like home. Whatever the Revolution's rhetoric of inclusion, Hassan found that race counted for more than the Qoran said it should. 'Here,' he said, 'you've got Iranian Muslims and that's fine and then you've got non-Iranian Muslims and they're a step down.'

'So you ended up feeling the same in Iran as you did in America?'

He nodded. 'Don't get me wrong. Being black is not the problem. The problem stems from not being Persian. Take the Arabs, the Turks, the Pakistanis; the Iranians don't think much of them either.'

He was smiling; his eyes had narrowed behind his round spectacles. 'I'm sure you've heard how the Iranians are the most intelligent people in the world?' I smiled back: ah yes, the Persian self-regard. Hassan was fed up with being told all the time how fine the Persian language was, how Persian culture and hospitality were second to none. He had a theory about Iranians and the monarchy. They couldn't get the ingrained servility out of their system, so the Islamic Republic had become a monarchy – with a king, a court and the same sycophancy and rivalries. I nodded. These thoughts had crossed my mind, as they cross the mind of many foreigners who spend time in Iran. But I was free to leave; Hassan was not.

He said, 'More important than feeling at home is the feeling that you're creatively useful.' It was some time since he'd had such a feeling. Nowadays, he was doing mostly freelance work, solitary stuff like translating and editing and the odd bit of teaching. He didn't feel like working in an office again; with their disloyalties and tensions, Iranian offices could be unhealthy places.

On Hassan's suggestion we ditched the Pride in Ghazvin. Despite the antiquity of its design, Hassan felt that a Paykan would cope better with the arduous drive over the mountains to Alamut. We found a Paykan and a driver, a talkative chap called Karimi. As we drove, I sensed that Hassan was mildly discomfited whenever Karimi addressed us a nosy question from the front seat, or interrupted our conversation to impart some piece of local knowledge. I knew that Hassan understood and spoke Persian, but he didn't care to use it. He pretended, in a semi-humorous way, that his Persian was worse than it was. Perhaps it was a way of underscoring, to himself and others, the temporary nature of his stay.

Each time détente had seemed possible between Iran and America, Hassan must have wondered if his case would be put on the negotiating table. Might the Iranians drop him in the Americans' lap, as part of a broader deal? He had hardly behaved in a way that would win support for leniency should he ever stand trial in the US. The *New Yorker* interview, for instance, had taken place in 2002, and he'd duly condemned the attack on the World Trade Center that had taken place a few months before. As he told me, 'The religion is very clear; you don't attack non-combatants.' But, he'd told his interviewer that the Pentagon – a military target – was fair game: 'I felt sorry for the people who died there, especially the civilians, but in a situation like that they knew where they were working. If I had to choose a target in Washington, it would be the White House.'

I didn't doubt that Hassan believed what he was saying, but his defiance also seemed like a personal pillar – a way of validating the momentous decision that he'd taken in 1980, to kill a man, and a means of dealing with the ramifications of that decision. Hassan told me that his lack of contrition had unnerved his American interviewers. 'I had the impression that they wanted me to say that I regretted it – it was as if they'd been personally offended.' As Hassan spoke, I was reminded of all the burned-out Iranian revolutionaries I knew. They were mellow and disappointed but I'd never heard one of them regret the things that he'd done. As Alavi Tabar had said, 'One can criticize the past, but one should never take revenge on it.'

He shared a second pillar with Alavi Tabar: religion. As far as I could tell, he was observant, though not ostentatiously. He said his prayers regularly and drank non-alcoholic beer. I wondered how his succession of temporary wives – his 'chequered career with Iranian women', as he put it – fitted into the religious frame. (Awkwardly, perhaps, for Sunni Islam – the Islam that Hassan embraced in 1969 – doesn't recognize the legitimacy of the Shi'a institution of temporary marriage.) But I sensed a will to preserve his faith, to keep it clean while the Islam of others was being spoiled. He was angry that religion had been exploited and used by the Islamic Republic, angry at the acts of despotism that were carried out in its name. He had never, he told me, suffered a personal crisis of faith, but he saw the irony that lay in the flight of normal Iranians from religion. 'They call this an Islamic Republic, but these guys are going to face an anti-religion backlash – it's starting already.'

I'd spent most of the journey bent over my notepad, recording what Hassan said as the Paykan strained up steep slopes and rounded hairpin bends. Now Hassan stopped talking and I looked out and saw that we were on a whale's-back ridge high above the valley floor; blueish mountains stretched away to the north. The air was buoyant and the sun was brilliant and suddenly I was overjoyed to be out of Tehran. As we made our descent, we saw harvested wheat shimmering in the valley below and villages made up of flat-roofed houses shaded by poplars and sycamores. Hassan said, 'If I had the means, I'd build a mud and brick house here, and fill it with modern accoutrements, and perhaps never come out.'

In a teahouse at the side of the road, I asked, 'Would you kill again?'

Since the Tabatabai hit he'd been approached twice by the Iranians but had rejected their proposals for logistical reasons. Once, they asked him to assassinate Saddam Hussein; recognizing the impossibility of the mission, he said no. In the late 1980s, he spent eighteen gruelling months fighting the Soviets alongside the Afghan Mujahedin. (The Soviets had put a $35,000 bounty on the head of any captured Western *mujahid*, so Hassan pretended to be a South African.) Referring to Islamic warriors, he said, 'The religion doesn't put restrictions on age', but I got the impression that his time of action had passed. Violence was for the young and fearless and Hassan had become middle aged.

He emphatically didn't repudiate the use of violence so long as he considered the cause to be right. Referring to barely reported American atrocities that were alleged to have taken place in Afghanistan and Iraq, he said, 'I get upset at the idea that someone sits in their air-conditioned office and orders the razing of some village without fear of retribution. Generally, I think there should be a better way than killing. But it makes all the sense in the world that those who kill should fear for their own lives.'

He regarded George Bush as the spiritual descendant of the Crusaders who fought against Islam. But he reserved the most venom for Condoleezza Rice and Colin Powell. It was as if he believed that they had forfeited their blackness. 'Rice,' he said, 'was the first to raise the subject of regime change in Iraq. She made a speech comparing war against Saddam Hussein to the civil rights movement in the US. That's BULLSHIT. It bothers me, you know. At fifty-two, it bothers me.' He shook his head. 'You know, the analogy with the Crusades isn't such a foolish one. The guys who were doing the fighting then were guys who hadn't had any contact with Muslims. It's the same now. The Americans are fighting something that they don't understand.'

We were drawing into a small village; I looked to the right and saw the rock of Alamut. It was grey and desolate and joined by a granite saddle to a smaller sandstone protrusion; Juvaini, a Persian chronicler of the Mongol period, likened this arrangement to a camel lying down with its neck stretched out. It's said that a local king chose this site for his castle after a manned eagle, released on his orders, landed on the rock. The district seems to have been renamed after this auspicious event, for Alamut meant 'eagle's teaching' in the language that was spoken locally at the time. Hassan told me he'd seen an eagle on the drive up, but I'd only seen crows with red-tipped wings, riding the currents.

To reach the steps that led up the northern face to the citadel we had to cross a stream that served the village. The stream had received no rainwater during the summer and its feeble flow had been partially diverted by some rocks. As we got out of the car, a girl of about fifteen appeared with a spade and started prising the rocks from the bed, biffing them out of the current. When she'd finished, and the flow was restored, she used the spade as a pole to vault agilely back across the stream. As she trotted back to the village we started climbing. After a few steps, Hassan turned around, panting. He said, 'I challenge you to find a fat man in these valleys.'

Recognizing the site's impregnability to physical assault, Hassan-i-Sabah had used guile to win it. He had secretly converted the garrison of soldiers within and had himself smuggled inside the walls. The *seigneur* was an appointee of the Seljuk state; he withdrew and was rewarded for his prudence, the chroniclers relate, with three thousand gold dinars. For the next thirty-five years, Hassan-i-Sabah and his devotees endured sieges and assaults while campaigning against the Seljuks, the Crusaders and, after a schism that took place in 1094, against the main branch of Ismailis, in Cairo. During this period, not only did he succeed in terrifying his opponents and winning secret converts, but he also spread his influence to a series of remote castles across Persia and the Levant. He was no longer personally a man of action; during his time at Alamut he is said to have left his house in the castle only twice. He spent much of his time committing his beliefs to the page. But his reputation for savagery and fanaticism overshadowed his erudition. When his emissaries brought the new preaching to Syria, which was being contested by the Seljuks and Crusaders, Western chroniclers fabricated a lurid demonology.

In his *Travels* of the late thirteenth century, Marco Polo writes that Hassan-i-Sabah 'had made in a valley between two mountains the biggest and most beautiful garden that was ever seen, planted with all the finest fruits in the world and containing the most splendid mansions and palaces that were ever seen, ornamented with gold and with likenesses of all that is beautiful on earth, and also four conduits, one flowing with wine, one with milk, one with honey, and one with water. There were fair ladies there and damsels, the loveliest in the world . . .' The lord of Alamut, Marco Polo goes on, would introduce his inductees into the garden, which they would naturally mistake for paradise, only to have them drugged and brought out. When they came to in the dank confines of the castle, they believed that Hassan-i-Sabah had the power to return them. 'Then, in order to bring about the death of the lord or other man which he desired, he would take some of these Assassins of his and send them wherever he might wish, telling them that he was minded to dispatch them to Paradise; they were to go accordingly and kill such and such a man; if they died on their mission, they would go there all the sooner. Those who received such a command obeyed it with a right good will.'

The truth was more prosaic. There were no mansions or flowing wine. For all the fanatical devotion of the Assassins to their lord, life at Alamut was overshadowed by intermittent fighting and negotiations with the Seljuks; one ruler sent his forces for eight successive years to destroy the Ismailis' crops and engage them in battle. Bernard Lewis, one of the foremost scholars of the Middle East, is not convinced by the claims of Marco Polo and many others that Hassan-i-Sabah administered hashish to his inductees. According to his Persian biographer, the lord of Alamut led an 'ascetic, abstemious and pious life'; he seems to have enjoined the same on his followers.

Halfway up the steps, Hassan and I rested on a platform built by the local authorities, in front of a lava ridge that was joined by a col to the rock. To the north and east, small cultivated valleys were visible, and tracks that led further into the range – tracks that were trodden by mules bearing rice from the Caspian provinces as recently as 1930, and as early as the second century, when they were mentioned by a Chinese chronicler.

A few minutes and three hundred steps later, we were on the granite bluff, about two hundred feet from the valley floor. 'My legs are heavy,' Hassan said; it was the first exercise he'd had in months. Near the southern edge of the rock there was what looked like the entrance to a cave. But the cave had no back wall; it was a massive porthole framing a stupendous view, of the village in the foreground and behind it the main Alamut valley, with the mountains we had crossed in the far distance.

Further up the rock, we found a woman from the cultural heritage department wearing a cap with a sunshade over her black head-cover. She was directing some local men who were digging, and got cross when I strayed into the excavation area. By the time the dig started, a few years ago, local people in need of building materials and bounty hunters looking for Hassan-i-Sabah's famous (and perhaps legendary) 'treasure' had destroyed much of what remained of the castle. The woman was standing at the lip of what she deduced had been a cistern. I asked about honey; there's a story about an enemy agent falling into a store chamber filled with honey, and drowning. The woman said that she and her men had come across plates and cups that had been imported from China. They dated from the Safavid period, when the castle was a gilded prison. When I asked about Hassan-i-Sabah's grapevine, she pointed to a tongue that extended from the western side of the rock. From our vantage point, looking into the ravine below, the vine seemed to sprout miraculously from the rock.

Perhaps this vine provided grapes for the wine that was drunk during the sovereignty of a second Hassan, the fourth Ismaili lord of Alamut. In 1164, Hassan proclaimed the millennium. This meant that holy law was no longer applicable; the chosen followers of the Imam had reached a state of grace that rendered conventional observance obsolete. Hassan allowed wine to be drunk, abolished formal prayers and punished those who persisted with the old laws. As far away as Syria, loyal Ismailis rejoiced at the new dispensation. According to one Sunni source, the inhabitants of one Levantine citadel 'gave way to iniquity and debauchery, and called themselves "the Pure". Men and women mingled in drinking sessions, no man abstained from his sister or daughter', and 'the women wore men's clothes'.

More than sixty years elapsed before a lord of Alamut called the people back to the law, and even made accommodations to orthodoxy. Then, in the mid-thirteenth century, the Ismailis faced their last enemy. By 1240, the Mongols had invaded western Iran and were making inroads into Armenia, Georgia and parts of Mesopotamia. Eighteen years later, they sacked Baghdad and put to death the Abbasid caliph, the titular head of Sunni Islam. Rukn al-Din, the head of Hassan-i-Sabah's order, tried to temporize in the face of Mongol demands that he surrender unconditionally. But his commander at Alamut capitulated and the Mongols destroyed the castle, including its famous library. And so the Ismailis faded. Juvaini wrote: 'Today . . . if an assassin still lingers in a corner he plies a woman's trade . . . the kings of the Greeks and Franks, who turned pale with fear of these accursed ones, and paid them tribute, and were not ashamed of that ignominy, now enjoy sweet slumber.'

In America, and the West in general, many people would regard Hassan Abdolrahman as an 'accursed one' – he, Hassan-i-Sabah and Ayatollah Khomeini together. Knowing Hassan slightly, I'm finding judgement to be as slippery as soap. From a lawyerly perspective, the legitimacy of Hassan's action in 1980 may indeed hinge on whether or not, as he says, this was war and he and Tabatabai were combatants. (In his postman's uniform, Hassan was an undeclared combatant; war or no war, the hit lacked gallan-try.) But I am not a lawyer and I have no cause. Travelling to Alamut with Hassan reminded me that there is something that prepares men not only to kill for their beliefs, but also to die for them. Men like David Belfield, like Ali Reza Alavi Tabar, Sadeq Zarif and Hossein Kharrazi had this thing, this whatever, this suicidal and homocidal vileness. They had its brilliant integrity, too.

CHAPTER TEN

Ashura

It was Ashura, the anniversary of the martyrdom of the Imam Hossein. A big crowd was gathered at an intersection, surrounding a blue conical tent. It represented the Imam's tent – the tent that his foes, following their victory, burned to the ground. I was in the crowd with a friend, Khosro. From our position, behind a police cordon that also encircled the tent, we heard the sound of chests being thumped, and the repeated intonation, 'Hossein!' We saw the mourners' feet shuffling under the bottom of the tent, which was three feet or so off the ground. These people must have got here early; the police weren't letting anyone else across. The only people they admitted were some officials, who doused the tent with petrol.

Between the cordon and the tent, an apathetic performance was in train. Shemr, Yazid's diabolical commander, was trundling around, terrorizing little boys in green Arab robes. (We took them to be the children in Hossein's entourage.) He blew a whistle and occasionally lashed out with a horsewhip, but not hard enough to hurt the boys. There was no sign of Shemr's commander, Hor, who defected to Hossein's ranks as the battle was about to begin. And there was no sign of the Imam.

The crowd were waiting for the call to prayer: Hossein's death knell. They were arguing over whether everyone should sit or stand. As long as the people in front stayed on their feet, those at the back couldn't see anything. But the concentration of bodies meant there was hardly any room to sit; you'd get elbows and knees sticking into you. So some sat, others stood and everyone grumbled. People continued to arrive and the sun got stronger. The thumping from the tent contributed to the sense of lethargy.

Then, as Shemr's corrupt yellow army pursued the innocents, action, of a sort. An old man broke through the cordon. His movements were blurred – by emotion, perhaps, or age – but he reached Shemr before anyone noticed, and struck him feebly on the chest. For a moment, Shemr seemed caught between fury and amusement. A policeman arrived and escorted the old man back to his place. Then we heard the call to prayer.

Khosro beckoned me through the awkwardly kneeling bodies, past the praying cops, and we crossed the intersection and entered the blue tent. It contained thirty or forty men. The atmosphere was tense, self-contained. They were thinking of the plain at Karbala, littered with the bodies of martyrs, and of the Imam's mutilated body. They had increased the violence of their intonations and the blows they brought down on their chests. Then, for no reason that was apparent to me, they started to move around the pole that held up the tent.

The rotation got faster. Khosro and I were caught up. The intonations had become shouts: 'Hossein! Hossein! Hossein!' You had to grab the shoulder of the person in front, otherwise you'd get flung out of the orbit. A hand emerged from the mêlée, trying to grab a long green ribbon that had been wrapped around the tent pole; it had, in this moment of elec-trifying intensity, become holy. The hand was swept away by the fury of our revolution, but others replaced it – five or six, tugging. Eventually, the ribbon ripped and the men tore it to shreds. A roar went up in the tent, a kind of ecstasy.

A sound like a sigh. I looked up and saw pustules of flame rising up the sides of the tent. One or two men fell over. They leaped to their feet, and returned; no one wanted to leave the whirl of devotion.

Then the tent cleaved and smoke rose from a wound into the sky and there was a pure moment when the intonations lost their form and the stringing together of the Imam's name became a formless cry of anguish and unexpressed desires. I saw one man clutching a baby who was fast asleep. The man was yelling and crying as he whirled.

A spiral of fire lifted the burning tent into the sky and suddenly there was no distinction between those in the tent and those outside. The pole stood, but the tent had become black snowflakes falling to earth. Some of the men sat or squatted, and wept. Others juggled charred fragments, blowing on them, but softly lest they fly away. Women rushed up to smear ash onto their chadors.

And then, miraculously, the people formed a procession, instinctively segregated. From the women's section, there came a sound I had never heard before, like cats crying.

BIBLIOGRAPHY

Adelkhah, Fariba, *Être Moderne en Iran*, Éditions Karthala, 1998

Amirsadeghi, Hossein (ed.), *Twentieth-Century Iran*, Heinemann, 1977

Avery, Peter, Hambly, Gavin and Melville, Charles (eds.), *The Cambridge History of Iran*, Volume 7, *From Nader Shah to the Islamic Republic*, Cambridge University Press, 1991

Bakhash, Shaul, *The Reign of the Ayatollahs: Iran and the Islamic Revolution*, IB Tauris, 1985

Brown, Ian, *Khomeini's Forgotten Sons: The Story of Iran's Boy Soldiers*, Grey Seal Books, 1990

Brumberg, Daniel, *Reinventing Khomeini: The Struggle for Reform in Iran*, University of Chicago Press, 2001

Bulloch, John and Morris, Harvey, *No Friends but the Mountains: The Tragic History of the Kurds*, Viking, 1992

Byron, Robert, *The Road to Oxiana*, Macmillan and Co., 1937

Cordesman, Anthony H. and Wagner, Abraham R., *The Lessons of Modern War*, Volume 2, *The Iran–Iraq War*, Westview, 1990

Daftary, Farhad, *The Assassin Legends: Myths of the Ismailis*, I.B. Tauris, 1995

Deniau, Jean-Charles, *Bani-Sadr: Le complot des ayatollahs*, Editions la Decouverte, 1989

Dieulafoy, Jane, *La Perse, la Chaldée et la Susiane*, Librairie Hachette et Cie, 1887

Ganji, Akbar, *Alijenab Sorkhpoosh va Alijenaban-e Khakestari*, Entesharat-e Tarh-e no, 1378, 1999–2000

—, *Tarikhane-ye Ashbah*, Entesharat-e Tarh-e no, 1378, 1999–2000

Halliday, Fred, *Iran: Dictatorship and Development*, Penguin, 1979

Hiro, Dilip, *Iran under the Ayatollahs*, Routledge and Kegan Paul, 1985

Jackson, Peter and Lockhart, Laurence (eds.), *The Cambridge History of Iran*, Volume 6, *The Timurid and Safavid Periods*, Cambridge University Press, 1986

Kadivar, Mohsen, *Hokumat-e Velayi*, Nashrani, 1377, 1998–1999

Kapuscinski, Ryszard, *Shah of Shahs*, Harcourt Brace Jovanovich, 1985

Katouzian, Homa, *Musaddiq and the Struggle for Power in Iran*, I.B. Tauris, 1999

—, *State and Society in Iran: The Eclipse of the Qajars and the Emergence of the Pahlavis*, I.B. Tauris, 2000

Khalkhali, Sadeq, *Khaterat-e Ayatollah Khalkhali*, Nashr-e Saye, 1379, 2000–2001

Khomeini, Ruhollah, *Hokumat-e Eslami*, Najaf, 1971

Khosrokhavar, Farhad and Roy, Olivier, *Iran: Comment sortir d'une révolution religieuse*, Éditions du Seuil, 1999

Kinzer, Stephen, *All the Shah's Men: An American Coup and the Roots of Middle East Terror*, John Wiley and Sons, Inc., 2003

Lewis, Bernard, *The Assassins: A Radical Sect in Islam*, Weidenfeld & Nicolson, 1967

Lewis, Franklin D., *Rumi – Past and Present, East and West*, Oneworld, 2004

Lowe, Jacques, Maisel, Jay and Glinn, Burt, *Celebration at Persepolis*, Creative Communications. n.d.

Mango, Andrew, *Ataturk*, John Murray, 1999

McDowell, David, *A Modern History of the Kurds*, I.B. Tauris, 1996

Milani, Abbas, *Mo'ama-ye Hoveida*, Nashr-e Akhtaran, 1380, 2001–2002

Mirzai, Sina, *Teyyeb dar Gozar-e Lootiha*, Entesharat-e Medya, 1381, 2002–2003

Moin, Baqer, *Khomeini: Life of the Ayatollah*, I.B. Tauris, 1999

Montazeri, Hossein-Ali, *Khaterat-e Ayatollah Montazeri*, Entesharat-e Enqelab-e Eslami, 1379, 2000–2001

Mottahadeh, Roy, *The Mantle of the Prophet: Religion and Politics in Iran*, Oneworld, 1985

Moussavi-Faridani, Mohammad-Ali, *Esfahan az Negahi-ye Digar*, Entesharat-e Naghsh-e Khorshid, 1378, 1999–2000

O'Balance, Edgar, *The Gulf War*, Brassey's Defence Publishers, 1988

Pahlavi, Mohammad Reza, *The Shah's Story*, Michael Joseph, 1980

Polo, Marco, *Travels* (trans. R. E. Latham), Penguin, 1958

Pope, Arthur Upham, *Introducing Persian Architecture*, Soroush Press, 1976

Randal, Jonathan C., *After Such Knowledge, What Forgiveness? My Encounters with Kurdistan*, Westview, 1999

Richard, Yann, *Shi'ite Islam*, Blackwell, 1995

Savory, Roger, *Iran under the Safavids*, Cambridge University Press, 1980

Sciolino, Elaine, *Persian Mirrors: The Elusive Face of Iran*, Free Press, 2000

Shirazi, Sayyad, *Khaterat-e Sepahbod Shahid Ali Sayyad Shirazi*, Sazman-e Tablighat-e Eslami, 1378, 1999–2000

Sick, Gary, *October Surprise: America's Hostages in Iran and the Election of Ronald Reagan*, I.B. Tauris, 1991

Stark, Freya, *The Valleys of the Assassins and Other Persian Travels*, Modern Library, 2001

Tehranchi, Mohammad Mehdi, *Varzeshhaye Zorkhaneh-e Iran*, Ketabsara Moqabel-e Daneshgah-e Tehran, 1374, 1995–1996

Timmerman, Kenneth R., *The Death Lobby: How the West Armed Iraq*, Houghton Mifflin, 1991

Torbati Sanjabi, Mahmoud, *Koudetasazan*, Mo'assesse-ye Farhang-e Kavush, 1376, 1997–1998

Wright, Denis, *The English Among the Persians*, I.B. Tauris, 2001

Wright, Robin, *In the Name of God: The Khomeini Decade*, Touchstone Books, 1990

Wroe, Ann, *Lives, Lies and the Iran–Contra Affair*, I.B. Tauris, 1991